Jan. 03

To: Dr. Stoddard

Rennie Beumuse

The New Science of Possibilities:
II. The Processing Technologies

by

Robert R. Carkhuff, Ph.D.
and
Bernard G. Berenson, Ph.D.
with
Christopher J. Carkhuff, M.A. Cert.

Contributors:
Alvin A. Cook, Ph.D.
Andrew H. Griffin, Ed.D.
Shirley McCune, D.S.W.
David C. Meyers, M.A.
William O'Brian, Ph.D.
Peter Rayson, M.S., C.Eng.
Darren Tisdale, M.A.

Published by Possibilities Publishing
22 Amherst Road
Amherst, MA 01002
413-253-3488
1-800-822-2801
413-253-3490 (fax)
www.possibilitiespublisher.com

ISBN 0-87425-590-2

Editorial services by Mary George
Production services by Jean Miller
Cover design by Donna Thibault-Wong

To Our Mentors . . .

B.R. Bugelski, Ph.D., who empowered us in the probabilities of parametric science

and

James Drasgow, Ph.D., who empowered us in the possibilities of non-parametric science

The New Science of Possibilities
II. The Processing Technologies

CONTENTS

About the Authors

DR. ROBERT R. CARKHUFF is Founder and Chairperson, The Carkhuff Group: Human Technology, Inc., Human Resource Development Press, Inc., and Carkhuff Thinking Systems, Inc. One of the most frequently referenced writers in the social sciences, Dr. Carkhuff has authored many major works, including *Human Processing & Human Productivity, Sources of Human Productivity, The Exemplar, Empowering,* and *The Age of the New Capitalism.*

DR. BERNARD G. BERENSON is Executive Director, Carkhuff Institute of Applied Science and Human Technology. A professor at the state universities of Maryland, Massachusetts, and New York, Dr. Berenson was co-founder of the original Center for Human Resource Development and chairperson of its graduate training program at American International College. He is the author of *Sources of Growth in Counseling and Psychotherapy* and *Beyond Counseling and Therapy,* as well as contributing author to *The Development of Human Resources, Helping and Human Relations,* and many other books.

Together, Carkhuff and Berenson have introduced several market revolutions over the past four decades. They brought us the first interpersonal skills system in the '60s and '70s. They introduced the first individual models of human resource development in the '70s and '80s. Their combined efforts also initiated us into twenty-first-century thinking *in the '80s* with the first models of new leadership and organizational capital development. Recently they co-authored *The Possibilities Leader* and *The Possibilities Organization,* both applications of their new science of possibilities management. Now their longtime collaboration has reached yet another pinnacle of achievement with the two volumes of *The New Science of Possibilities.*

About the Co-Author and Contributors

CHRISTOPHER CARKHUFF is Director, Research and Development, Carkhuff Thinking Systems, Inc. Our co-author, he developed many of the research projects presented in this volume.

ALVIN A. COOK is President, Planning Systems International, Inc. He is a contributor to paradigmetrics, discussed in Chapter 7, Concurrency and Paradigmetric Technologies.

ANDREW H. GRIFFIN is Assistant Superintendent for Higher Education, Certification and Community Outreach, Washington. He is a contributor to *"The Possibilities Schools"* and *"The Possibilities Community"* (Chapter 7).

SHIRLEY MCCUNE is Director, Multi-States LINKS Project, and a contributor to *"The Possibilities Schools"* and *"The Possibilities Community"* (Chapter 7).

DAVID C. MEYERS is Director, Organizational Capital Development, Human Technology, Inc. He is a contributor to the SALT systems, discussed in Chapter 2, DNA' and Inductive Processing.

BILL O'BRIEN is Director, Center of Excellence, The Parametric Technology Corporation, Inc., and a contributor to paradigmetrics (Chapter 7).

PETER RAYSON is Senior Vice President, Industry Group, The Parametric Technology Corporation, Inc., and a contributor to paradigmetrics (Chapter 7). He is also a contributor to hybrid modeling, discussed in Chapter 5, Changeability and Hybrid Modeling.

DARREN TISDALE is Chief Technology Officer, Document Forum, Inc., and a contributor to paradigmetrics (Chapter 7).

Foreword

The Possibilities Economics

by Peter Rayson, M.S., C.Eng.
Senior Vice President, Industry Group
The Parametric Technology Corp., Inc.

It is common today to write about the decline and fall of
*"probabilities economics."** To be sure, its promulgators can no
longer predict what they cannot control: they simply cannot control
the implications of the trillions of decisions made in the marketplace each day.

What is needed is the vision of a new science of possibilities:
"the possibilities economics." Just as probabilities economics narrowed our focus to yield a startling array of goods and services in
the twentieth century, so will possibilities economics generate a
spiraling array of *"breakthrough"* products and services in the
twenty-first century.

The difference between possibilities and probabilities economics is not academic. It is the difference between expanding infinitely
and narrowing infinitesimally: the systematic expansion of possibilities economics generates an infinite array of *virtual* products,
services, and solutions. The systematic narrowing of probabilities
economics produces infinitesimal deviation or variability from
artificially imposed measures of central tendency.

The contrast between possibilities and probabilities is seen
most vividly in their differentiated processing and planning systems. Probabilities emphasize planning; planning is an exclusive
process that narrows tolerance through methods like statistical
process control. It is a necessary condition for producing a product.
It is an insufficient condition for producing the best products and
services: it requires the generativity of possibilities processing.

Possibilities, in turn, emphasize processing: generative processing that censors legacy learning; interdependent processing that
makes us *"one"* with our phenomena; continuous processing that
orients us to the changeable conditions of our environment. Possi

---

* Karlgaard, R. "The Four Mice of the Apocalypse." *Forbes,* September 20, 1999,
p. 45.

bilities processing is necessary for the generation of the best products and services. It is insufficient for producing these goods and
services: it requires the planning and production of probabilities.

While the planning systems of probabilities exclude the
processing systems of possibilities, the possibilities systems seek
only to drive probabilities. Probabilities ensure that we *do the
things right*. Possibilities ensure that we *do the right thing!*

The real differences between possibilities and probabilities
economics lie in the sciences that undergird them. Probabilities
science has as its historic functions description, prediction, and
control. Its problem is with spiraling change. In order to make
probabilities science work in the twenty-first century, its propo-
nents must re-order its functions: description, control, prediction.
This definition points up the absurdity of the science in a world of
changing conditions: *"Ready, fire, aim!"*

Possibilities science acknowledges changing conditions. It seeks
(1) to align itself with all phenomena by *relating* to phenomena, (2)
to intervene to *empower* or *enhance* the phenomena's potential, and
(3) to *release* or *free* the phenomena to make their own unique contributions and, in so doing, seek their own changeable destinies.
Possibilities science operates by the simplest principle: *"What you
cannot control, release!"* Since we cannot control anything, this
means unleasing everything: *Relate, empower, free!*

Indeed, the possibilities economics is more than just the source
of probabilities. It has its own continuously generated source of new
capital. Assuming that capitalism is merely a theory of change, it
deductively generates sources of new capital development:

- MCD—Marketplace capital development, or corporate
  positioning in the marketplace;

- OCD—Organizational capital development, or corporate
  alignment with positioning;

- HCD—Human capital development, or human processing
  to make alignment work;

- ICD—Information capital development, or information
  modeling to operationalize human processing;

- mCD—Mechanical capital development, or mechanical
  tooling to implement information modeling.

Together, these new capital development systems source the 90 percent of economic productivity growth not accounted for by financial capital.

In *The New Science of Possibilities,* Carkhuff and Berenson introduce not only the processing science and its technologies, but also the building blocks of *The New Economics of Possibilities.* Indeed, they introduce us to *The New Culture of Possibilities.*

Foreword

by Shirley McCune, D.S.W.
Director, Multi-State LINKS Project
Office of Superintendent of Public
Instruction, State of Washington

Science is a spiritual quest. Whether metaphysical or mathematical, the search is for meaning in our universe. Whether probabilistic and parametrical or possibilistic and nonparametrical, the search for meaning is addressed with rigor.

In introducing the new science of possibilities, Carkhuff and Berenson address both meaning and rigor at the very highest levels. The meaning of possibilities science is found in the functions of their possibilities science model:

- *Relating* to phenomenal experience;

- *Empowering* phenomenal potential;

- *Freeing* the phenomenal to search out their own unique and changeable contributions.

The rigor of possibilities science is found in the means these scientists employ to accomplish the functions of phenomenal relating, empowering, and freeing. The possibilities science model makes one fundamental assumption: everything, every dimension of everything, is processing. The implications of this assumption are profound: the actualization of all phenomena is found in the exploding possibilities of interdependency within, between, and among these processing dimensions.

The principles of possibilities science are derived from this fundamental assumption of *process-centricity*.

- All phenomenal processing systems are *"nested"* in higher-order processing systems.

- All phenomenal processing systems are rotated to become higher-order processing systems.

- All phenomenal processing systems are impregnated by the socio-genetic coding of higher-order processing systems and, in turn, transmit their genetic coding to lower-order processing systems.

xvii

These principles enable our scientists to construct the basic model for possibilities science:

> ***Relating, empowering, and freeing functions are discharged by phenomenal information-modeling components enabled by phenomenal interdependent processing systems.***

Again, all dimensions are processing dimensions and all dimensions are related interdependently.

Carkhuff and Berenson are simply and profoundly building upon the platform created by Einstein: *"It's all relative!"* For them, it's all about relating:

- All phenomenal processing systems relate interdependently, both internally and externally.

- All phenomenal processing systems rotate to ascendancy as driving functions in interdependent phenomenal relating.

- While all phenomenal processing systems relate interdependently, interdependent processing is not possible without the *"inequality"* of the phenomena.

In other words, while everyone and everything gets its turn, none are equal in the relative processing potential.

Clearly, what is most heuristic in the work of these *"scientist-biographers"* is their science of phenomenology. Dedicated to releasing phenomena to fulfill their changing potential, they build their phenomenal information models developmentally:

- Conceptual information,

- Operational information,

- Dimensional information,

- Vectorial information,

- Phenomenal information.

They then employ these models deductively to facilitate hypothesis-testing: phenomenal, vectorial, dimensional, operational, conceptual. This resulting modeling is the accelerant for evolutionary movement in civilization.

These *"scientist-biographers"* do not seek the balance of probabilistic approaches to interdependent phenomenal processing. Their phenomenal images are *not* linear, independent, symmetrical, static, and planning-centric. Instead, their phenomenal models are multidimensional, interdependent, asymmetrical, changeable, and process-centric. In other words, they are not committed to what is known; they are committed to probing the unknown to make it *"knowable."*

For our *"scientist-biographers,"* the catalysts or enabling processes are always human:

- Information relating,

- Information representing,

- Individual processing,

- Interpersonal processing,

- Interdependent processing.

These human processing systems are identified as *"I⁵."* The highest level is interdependent phenomenal processing, where the possibilities scientists process interdependently for and with the phenomena they have modeled.

Drawing upon the *"I⁵"* human-processing dimensions, we have modeled *"The New 3Rs"* learning-skills technologies:

- *Relating* to information operations,

- *Representing* informational operations,

- *Reasoning* with new and more powerful information representation.

The New 3Rs enable our learners to process creatively. The New 3Rs leverage the old 3Rs to improve all learning skills. The New 3Rs empower our learners to be prepared to meet the requirements of the twenty-first-century global marketplace.

In short, the work of these possibilities scientists will empower all of us—learners and workers, teachers and managers—to achieve *"equity and inequality"* through relating, empowering, and freeing the phenomena—people, data, things—to which we are dedicated.

For untold generations, we will owe these scientists a debt of gratitude for the definition of the platform for possibilities science:

a platform upon which past geniuses unknowingly built; a platform upon which future generations of rank-and-file geniuses will build.

In summary, what the possibilities science model does is allow us to generate powerful and accurate images of the phenomena we are addressing, images of their processing systems. Moreover, the model empowers us to process interdependently with these images to generate still more powerful and accurate images of the changing phenomena.

Armed with such possibilistic phenomenal images, we may generate our probabilistic designs and plan our programmatic outcomes. You see, the possibilities scientists do not ask us to abandon our probabilistic technologies. Rather, they treat probabilities science as they do any other phenomena: relating to it to experience its potential; empowering it to enhance its potential; releasing it to realize its potential.

Indeed, truth-seeking probabilities science is not possible without possibilities science. To be sure, possibilities science is the source of all probabilistic outcomes. Possibilities science is the spiritual quest! These scientists make it valuable to all of us in the new science of possibilities.

The Preliminaries . . .

Phenomena are the "stuff " of reality—entities that present themselves to our senses and understanding. People, data, things are all phenomena; even images we generate from such entities are phenomena. We can say that, potentially, the interrelationships within, between, and among the dimensions of these entities are also phenomena. And we must be aware that our senses and understanding are themselves phenomena. Science takes phenomena as its subject and begins with an application of vision; consequently, the vision we bring to our science is a crucial factor in the sufficiency of that science.

The science of probabilities is parametric science: it defines phenomena by their deviation or variability from some standard or central tendency. Although still service-able, probabilities science no longer suffices for defining phenomena, which are inherently changeable. We see this insufficiency in the context of its functions:

- *Describing* only small segments of variability due to limited linear measurement and extraordinary errors in the perception of phenomenal nature;

- *Predicting* only fictional results due to artifactual data produced by artificial measurement;

- *Controlling* only "two-response" conditioned mentalities due to artificially repetitious measurement.

In summary, lacking a true paradigm, probabilities science produces a collective body of independent and socially isolated categories of knowledge, thus trivializing the processing power of all phenomena.

The new science of possibilities is paradigmetric science: it defines continuously changing phenomena by continuously generating new paradigms and measurements for those paradigms. Possibilities science becomes an integral part of phenomenal process-centricity through these core functions:

- *Relating* empathically to all phenomenal experience;

- *Empowering* to enhance all phenomenal potential;

- *Freeing* phenomenal potential to discover its own changeable destiny.

In summary, possibilities science is a paradigm for generating and measuring paradigms. It is a new science that actualizes the socio-genetic, process-centricity of all phenomena.

Preface

The New Science of
Information Capital Development

The New Science of Possibilities empowers us with a whole new vision of possibilities for ourselves and our universes. Finally, we can learn to relate interdependently to phenomenal information operations as they exist in reality (see Table 1). We can do this because we understand the following:

- Phenomenal functions emphasize processing, not content.
- Phenomenal components are multidimensional, not linear.
- Phenomenal processes are interdependent, not independent.
- Phenomenal conditions are asymmetrical, not symmetrical.
- Phenomenal standards are changeable, not static.

Clearly, phenomenal operations are not what we have conceived them to be. They are indeed the very essence of possibilities science—the information-capital components that discharge the

Table 1. A Comparison of the Phenomenal Operations of Probabilities and Possibilities Sciences

	SCIENCES	
OPERATIONS	PROBABILITIES	POSSIBILITIES
FUNCTIONS	Content	Processing
COMPONENTS	Linear	Multidimensional
PROCESSES	Independent	Interdependent
CONDITIONS	Symmetrical	Asymmetrical
STANDARDS	Static	Changeable

freeing phenomenal functions. In short, science is information capital development, and information capital development is phenomenal operations.

Clearly, phenomenal operations are not what we have conceived them to be. They are indeed the very essence of possibilities science—the information-capital components that discharge the freeing phenomenal functions. In short, science is information capital development, and information capital development is phenomenal operations.

By way of illustration, the whole area of information technology, or IT, deserves a new vision. A precondition of the mission of IT is, to be sure, connectivity: the unimpeded flow of global information. However, the IT mission does not culminate with "nonsense syllables"; it culminates with the information modeling of phenomenal operations.

The data sent out over the Internet and other vehicles do not quality as information:

- Linear representations of phenomena are useless. Even matrices are inadequate because we do not get to see the phenomenal interactions in perspective.

- Independent representations of phenomena are artifactual. Nothing in this world is independent of anything else.

- Symmetrical curvilinear representations of phenomena are contrived. No phenomena are normally distributed, least of all humans.

- Static representations of phenomena are attenuated. Increasingly, our artificial curves shrink to become leptokurtic before disappearing.

- Planning-centric representations of phenomena are impotent. Without processing, planning accounts for the implementation, not the generation, of phenomena— perhaps less than five percent of phenomenal reality.

Only models representing phenomena in their full dimensionality qualify as information:

- Information is multidimensional. We cannot view phenomena in full perspective without multidimensional images.

- Information is interdependently related. Phenomena relate not only interactively but also interdependently through processing.

- Information is asymmetrical. All phenomena are asymmetrically curvilinear. Just look at the patterns of globalization or the Internet itself.

- Information is changeable. All phenomena are continuously changing.

- Information is process-centric. Socio-genetic process-centricity is the order in the universe.

Such phenomenal information modeling is the central concern of this book: phenomenal information modeling relates, empowers, and frees phenomena to discover their own changeable destinies. In essence, possibilities science is continuous and interdependent information modeling dedicated to continuous and interdependent changing phenomena.

For example, the recent focus of business upon inter-enterprise relationships creates the requirements for a common deductive and interdependent information environment. Within this information environment, entire constellations of business organizations inter-relate dynamically to redescribe themselves. Neither the businesses nor the information technologies employed by them will actualize their contributions until they are interdependently synergistic with human processing.

In this context, case studies in Chapters 4 through 7 present interdependent common object modeling, which will serve the new organizing function: processing for concurrent organization modeling and process-alignment services and software. These objects will **not** be probabilities objects. They will be *"possibilities objects,"* connected by the phenomenal operations of possibilities science, thus fulfilling Einstein's most fervent hopes.

We should know that in possibilities science, a new ideational movement is required to culminate the contributions of an old ideational movement. Just as industrial machinery actualized the agrarian movement, and information technologies culminated the mechanical contributions, so will the human technologies realize the contributions of IT. In turn, the organizational technologies will actualize the human technologies, and the marketplace technologies will realize the organizational technologies. Moreover, all technologies in interdependent and synergistic partnership will culminate the possibilities of the global community and its marketplace!

In short, the Age of Ideation will actualize the contributions of the Age of Information. Relating interdependently within, between, and among cultures and peoples will culminate our historic efforts for independence; free entrepreneurial spirits will realize the benefits of a new capitalism; direct democratic governance will actualize the contributions of an enlightened citizenry; all will be empowered by a new science of possibilities.

For science is nothing more, or less, than information capital development: phenomenal information components dedicated to relating, empowering, and freeing functions; interdependent processing systems enabling information capital development. That is the story of the possibilities scientist: interdependent processors dedicated to information capital development. Indeed, for possibilities scientists, the new science of possibilities is the interdependent and synergistic processing partner with the information capital that the scientists generate.

Welcome To
A Special Kind of Book!

This is a book of pictures. We call the pictures *information models*. Information models are the universal language of processing in the twenty-first century. They are the interdependent and synergistic processing partners with human processors or human capital. We relate to phenomena, represent them, and then generate increasingly more powerful information models of the phenomena. That is the thesis of this book: interdependent processing generates prepotent models of phenomena. The reader need not conquer the processing operations involved: there is time for that later with different experiences and materials. The reader need only experience the magical images of any and all phenomena; to be introduced to the generation of powerful possibilities; to realize the accomplishments of generations of work in an afternoon of processing.

RRC and BGB

I am convinced that someone will eventually come up with a theory whose objects, connected by laws, are not probabilities . . .

— Albert Einstein
Letter to Born

I

Introduction and Overview

An introduction to phenomenal possibilities:
multidimensional, interdependent, asymmetrical,
changeable, process-centric.

1 The New Science of Phenomenal Possibilities

When the Judaic-Christian theologies gifted us with *"a knowable God,"* they provided the platform for science: *the explication of the unknown.* The more we could probe the unknown, the closer we could get to knowing God.

Thus, science is inherently a spiritual quest. It is a search for the *soul* of the universe. It is a search for the *souls* of the humans who occupy some small space within that universe.

Science is a search carried out by these most humble of God's creatures, seemingly made in His image. It is a search carried on by the greatest of God's gifts—human brainpower, with its infinite capacity to generate and organize itself to capture and fulfill God's universes.

Science is the search that addresses the nature of nature. Indeed, the nature of nature is precisely the same as the nature of the scientist-biographers who pursue its secrets.

The nature of nature is social. The functions of nature are to relate: all of its systems relate interdependently within, between, and among themselves.

The nature of nature is informational. The components of nature are informational: all of its systems communicate informationally within, between, and among themselves.

The nature of nature is processing—interdependent processing: all of its systems process interdependently within, between, and among themselves.

Indeed, all of nature's systems are processing systems. It is only when the scientist-biographers utilize brainpower systems to generate images of nature's systems that the scientists and nature become *one.* It is only when the scientist-biographers process interdependently within, between, and among nature's systems that we come close to knowing God.

PROBABILITIES SCIENCE

The first of human efforts to formulate scientific systems to study phenomena may be termed *"probabilities science."* Probabilities science enables us to *control* phenomena within a small *window* of opportunity in space and time. *"Control"* is the operative word here, for the functions of all probabilities science culminate in control: we describe phenomena in order to predict them; we predict phenomena in order to control them for humankind's purposes. These are the fundamental functions of probabilities science: describing, predicting, controlling.

The assumptions of probabilities science enable us to accomplish this *control*. The first of these is the linearity of phenomena. Scientists assume that, in light of human conditioning, there is no real difference between human processing and the linear processing of mechanical systems; indeed, for many, humans continue to be modeled upon linear mechanical systems (see Figure 1-1). Even the branching systems of information technologies are based upon "*Go—No Go*" choices of linear systems: the discriminative learning systems that enable choice empower us to choose only between one linear system and another. In other words, *truth* is a *line:* a scale or measurement or program to achieve an objective.

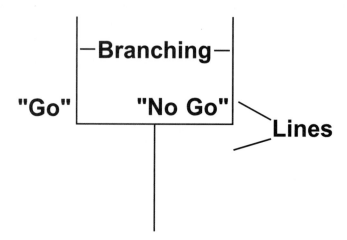

**Figure 1-1. The Probabilities Assumption
of Linearity**

The second of these assumptions of probabilities science is the independence of phenomenal factors in the universe. Independence is assumed in orthogonal terms: one factor is as likely to relate to another factor as it is likely *not* to relate at all (see Figure 1-2). In other words, the relationship between independent factors is random.

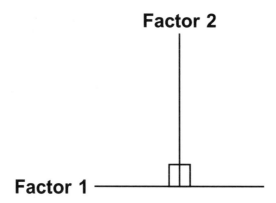

**Figure 1-2. The Probabilities Assumption
of Independence**

The third of these probabilities assumptions is the symmetry of the phenomenal curves. Symmetry is assumed in terms of the *normal* distribution of relationships between and among independent factors; even skewed curves are treated as *normal* curves in terms of the analyses of their central tendencies and variabilities (see Figure 1-3).

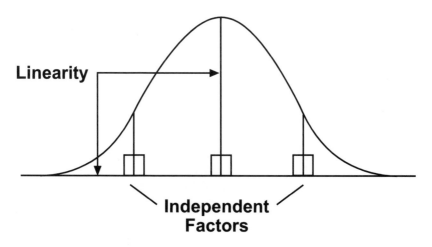

**Figure 1-3. The Probabilities Assumption
of Symmetry**

The fourth assumption of probabilities science is the constancy or stability of these independent factors. Constancy is assumed in replicable terms: because the variability of factors is narrowed, the factors remain the same; even when we do secondary and tertiary analyses, we analyze the same data we used during our primary analysis, which yielded our original factors (see Figure 1-4). In other words, the phenomenal factors and their independent relationships are enduring; the phenomena are constant.

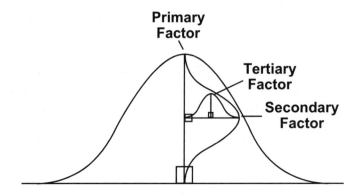

**Figure 1-4. The Probabilities Assumption
of Constancy**

The fifth assumption of probabilities science is the source of constancy: content-centricity, the continuous emphasis upon specific content areas and levels. Content-centricity is the limiting principle of all probabilities science: the focus upon linear, independent, symmetrical, and static representations of specific content areas. In this context, not only the dimensions but also the content are exclusive and isolated and, therefore, doomed to attenuation by spiraling changes in the environment.

The implications of these assumptions of probabilities science are profound. The probabilities analyses result in local probabilities of static, symmetrical, and independent linear phenomenal factors. These phenomenal probabilities are due to artifactual data caused by the loss of variance resulting from artificial *controls*. We may term these phenomenal probabilities *"node-centric"*: their window of opportunity in space and time reflects only a data point in the expanding network of phenomena.

POSSIBILITIES SCIENCE

The next of human efforts to scientifically process phenomenal experience may be termed *"possibilities science."* Possibilities science enables us to release or to free phenomena. The operative word here is *"free"*: the freedom of all phenomena, including human, is a function of the phenomena's processing systems; we relate to phenomena in order to comprehend their processing potential; we intervene to empower phenomena and thereby enhance their processing potential; we free phenomena in order to release their processing possibilities within God's universe of phenomenal possibilities.

The assumptions of possibilities science enable us to generate this freedom. The first of these assumptions is the multidimensionality of phenomena. Multidimensionality is assumed in the dimensions of the phenomenal processing systems: component inputs, transforming processes, function outputs, driving conditions, measurable standards (see Figure 1-5). In other words, phenomena, whatever their form, are inherently multidimensional.

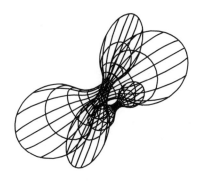

**Figure 1-5. The Possibilities Assumption
of Multidimensionality**

The second of these possibilities assumptions is the interdependence of phenomenal vectors in the universe. Interdependence is assumed in vectored terms: this does not mean mutual dependency, but partnered processing for potentially mutual benefits; partnered processing between scientists and phenomena; partnered, virtual processing between phenomena and phenomena (see Figure 1-6). In other words, the relationships between interdependent vectors are interactive and synergistic.

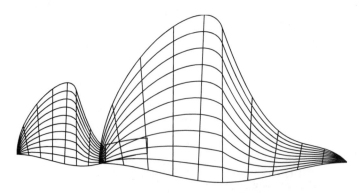

**Figure 1-6. The Possibilities Assumption
of Interdependence**

The third of these assumptions of possibilities science is the asymmetry of the phenomenal curves. When we work with the data excluded from probabilities analyses (which assume independent factors and normal distributions), we find that asymmetrical curves define the essence of the phenomena (see Figure 1-7). We thus assume asymmetrical models of the changing nature of phenomena: there is no base line from which the phenomena vary; only infinite and asymmetrical changeability.

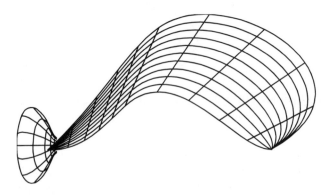

**Figure 1-7. The Possibilities Assumption
of Asymmetry**

The fourth assumption of possibilities science is the dynamic and continuous changeability of these interdependent vectors. Changeability is assumed in terms of 360 degrees of global freedom or diversity: due to the continuous expansion of the changeability of *all* vectors, the changeability of *any* vector is continuous (see Figure 1-8). In other words, only the continuously changing phenomena are enduring: the only constancy is change.

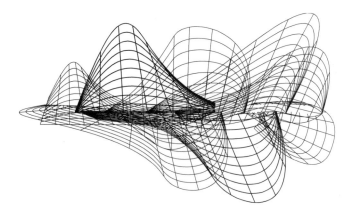

**Figure 1-8. The Possibilities Assumption
of Changeability**

The fifth assumption of possibilities science is the source of changeability: process-centricity, the continuous processing of all phenomenal dimensions. Process-centricity is the generating principle of all possibilities science: the multidimensional, interdependent, asymmetrical, and changeable processing of all dimensions. In this context, not only the dimensions but also the processing systems themselves are continuously changing.

The implications of these assumptions of possibilities science are profound. Possibilities science generates expanding global possibilities of continuously changing, interdependent, and asymmetrical multidimensional phenomenal vectors. These phenomenal possibilities are due to the processing ability to align with the phenomena in their naturalistic form. We may regard these phenomenal possibilities as process-centric in a potentially infinite web of networks reflective of God's universes of changing phenomena.

11

We may view possibilities and probabilities phenomena in perspective in Figure 1-9. As can be seen, probabilities phenomena occupy a small window of opportunity in space and time. Indeed, we may think of them as *"probabilities moments"* rather than as phenomena in themselves. They occur within infinite possibilities phenomena. This is the essence of possibilities science: probabilities moments occurring within the context of infinite possibilities; possibilities phenomena ensuring that we have accurate phenomenal perspectives of probabilities.

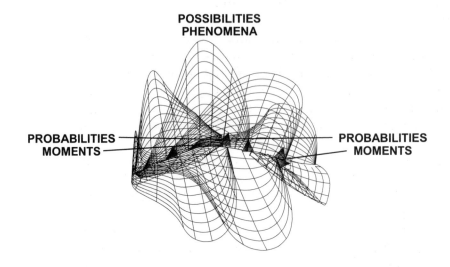

Figure 1-9. Probabilities Phenomena in a Possibilities Phenomenal Context

When we are producing products and services of any nature—whether they involve people, data, or things—we employ possibilities to drive probabilities: possibilities ensure that we are *doing the right thing.* We employ probabilities to produce our products: probabilities ensure that we *do things right.*

In turn, our possibilities design incorporates probabilities. Surely, we may employ probabilities, such as our knowledge of *best practices* in specific areas, to empower scientists to generate possibilities designs. In this instance, our probabilities empowerment enables our possibilities design.

To introduce possibilities models representing phenomenal possibilities, we may employ probabilities imagery: linearity, right angles, singular curvilinear tracks, and the like. However, the asymmetrical nature of phenomenal possibilities provides the framework for empowering probabilities moments: whatever the nature of phenomenal processing, the phenomena will reconfigure asymmetrically. *Thus, we can now freely employ free dynamic models to provide perspectives for static models:* whatever the nature of the phenomena, they will change continuously—birthing, growing, dying, rebirthing—to discover and rediscover their own changeable forms.

We also emphasize information models of all phenomena, doing so because we can understand them most readily. That is the nature of possibilities science. If information lives, then everything lives, for we can generate information models for everything. In this context, all science is applied science: *everything lives!*

While these sciences and the phenomena they generate are interdependent and synergistically related, the power of one dominates the contributions of the other: possibilities are expanded, infinite; probabilities are reduced, infinitesimal. Possibilities have been God's province. Probabilities have been, until now, humankind's province.

Although humans can be justly proud of their probabilities contributions, those contributions are infinitesimal in relation to our infinite phenomenal universes. The entire history of civilization, for example, has revolved around energy: discovering, extracting, refining finite resources of fossil fuels—coal and shale, oil and gas. Indeed, in the twentieth century, we fought two great world wars primarily over energy, killing more than 100 million people and destroying the lives of hundreds of millions of others.

Now, drawing from the generativity of the fictional *"Star Trek,"* we employ ion-driven, solar electric propulsion systems to thrust our spacecrafts into the heavens. Someday, burning fossil fuels will simply be a memory trace for civilization: a *probabilities moment.* What kind of a civilization will we create when we learn to align, enhance, and release the infinite phenomenal resources of God's multiple universes?

No, the describing, predicting, and controlling functions of probabilities science are not going to define and actualize our brave, new, and prosperous world; neither are the parametric assumptions, planning paradigms, and statistical process controls!

Yes, the relating, empowering, and freeing functions of possibilities science are going to generate and innovate our continuously growing and changing human and phenomenal destinies! Our unfettered, paradigmatic assumptions and process-centricity are going to release the power of the universes to literally and physically make everything out of nothing but our brainpower and our information models: our precious science of possibilities that drives our technologies of probabilities.

Clearly, in possibilities science, we serve a freeing function. The power of possibilities science is found in the interdependent processing of matured scientists with the phenomena they are addressing. The interdependent processors discern and apply the powerful forces guiding our universes: multidimensionality, interdependence, asymmetry, changeability. Both processor and phenomena mature!

Possibilities science is both source and force; content and processing; means and ends of all possibilities. Only our egos and their illusion of independence prevent us from understanding the interdependent nature of the universe.

We may represent these phenomenal possibilities symbolically by continuously changing, multidimensional, interdependently related, and asymmetrical curvilinear dimensions (see Figure 1-10). Indeed, just as phenomenal possibilities are changing, they are also expanding: quantitatively in the direction and force of their vectors; qualitatively in their increasingly inclusive power to create space and time. They are like an all-powerful changeable crystal, constantly exploding in curvilinear space with new and interdependently related, multidimensional systems: each yields the infinite possibilities of the changeable wholeness of our multiple universes.

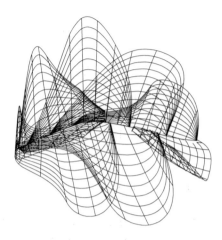

Figure 1-10. Generating Phenomenal Possibilities

As these universes of possibilities expand, they generate *probabilities moments:* opportunities upon which probabilities science may capitalize. As these universes of possibilities contract, they *become* probabilities phenomena. In this context, with the perspective of temporal and spatial distance, we may collapse this multidimensionality in singularity (see Figure 1-11). However, nature, having no tolerance for singularity, again explodes to generate new and changeable and expanding phenomena.

●

Figure 1-11. Singularity

In summary, the culminating implications of probabilities and possibilities sciences contrast vividly: infinitesimal to infinity! Given the constancy of linear, independent, and symmetrical phenomena, probabilities science is planning-centric. In contrast, given the changeability of multidimensional, interdependent, and asymmetrical phenomena, possibilities science is process-centric. Only continuous processing can align, empower, and free the perpetual motion of changing phenomena.

The fundamental principles of possibilities science thus empower us to generate infinite possibilities that both embrace and resolve the apparent conflicts of relativity and chaos: multidimensionality, interdependence, asymmetrical curvilinearity, and changeability. We can neither explore our universes nor penetrate infinity with the simplicity of the equations of probabilities science: $x = y$.

Indeed, the models of possibilities science that we have represented are highly initiative and, as such, have limitations even as they move inexorably toward discovering the nature of nature and its phenomena: multidimensional, interdependent, asymmetrical, changeable. This is *"Humankind's Work"!*

The phenomenal models, in turn, require a higher level of alignment through relating; of enhancing phenomenal potential through empowerment; of freeing phenomena to discover their own changeable destinies. They are therefore experiential, evolving, and unifying; they express the changing basic fabric of developmental merging toward singularity and the resulting explosion of infinite possibilities. This is *"God's Work"!*

Concurrency: A Story of Continuous Interdependent Processing

The paradigm for probabilities science is planning. We decide upon our goals and develop systems and programs to achieve them. The focus of our systems and programs is to most closely approximate the goals we have defined. Essentially, we plan in order to achieve our goals at the highest levels. We design our systems and develop our programs in a linear, or serial, manner: the parties of all operations deliver the products of their planning to the next parties in line.

The paradigm for possibilities science is processing: continuous interdependent processing; concurrent interdependent processing. We label this process-centricity *"concurrency"* because it is the true currency of processing. In concurrency, all parties of all operations process simultaneously; all parties of all operations receive the processing bulletins of all other parties and operations.

We may view the familiar probabilities planning paradigm in sharp relief in Figure 1-12. We initiate our systems design with an operational definition of our goals. As may be noted, the operations move systematically toward achieving the goals: resource inputs are converted into results outputs by transforming processes; the feedback loop measures the level of goal achievement and feeds it back as data input in a cyclical attempt to approximate the original goals.

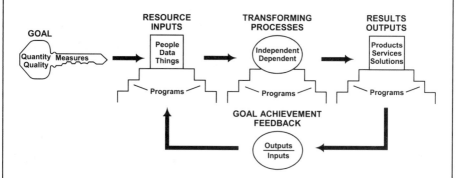

Figure 1-12. Probabilities Planning System

Analogously, the objectives within the inputs, processes, and outputs are defined as objectives, and programmatic step-by-step procedures are designed to achieve them: improved inputs, improved processes, improved outputs. Thus, the people, data, and thing inputs are defined operationally as objectives. Similarly, the independent and dependent processes are defined operationally as objectives. Likewise, the product, service, and solution outputs are defined operationally as objectives. The goals are static and never-changing. The programs may be conceived of as mini-systems designed to improve the operations. The goals are static.

A simpler view of the probabilities planning paradigm is presented in Figure 1-13. As depicted, we begin by defining the goals. Next, we design the systems that we need to achieve the goals: outputs, inputs, processes, feedback. We conclude by achieving some approximation of the goals—the same goals that we defined initially.

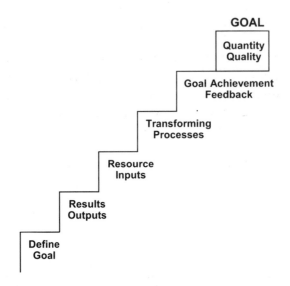

Figure 1-13. Probabilities Planning Paradigm

At Carkhuff Thinking Systems, we designed a process for concurrency: continuous interdependent processing.* In concurrency, we do not begin with an operationally defined goal; all parties process simultaneously and interdependently toward an undefined end. At the culmination of concurrency, we have generated qualitatively new opportunities: elevated inputs and outputs; elevated feedback and goal achievement. We have not processed these benefits independently and randomly; we have generated them interdependently and concurrently.

---

* Carkhuff, C. J., and Carkhuff, R. R. *Concurrency.* McLean, VA: Carkhuff Thinking Systems, Inc.

One of our missions was to employ our breakthrough visions of science and our innovative technologies to generate new businesses. As illustrated in Table 1-1, we employed an initiative GICCA processing paradigm to represent the components of the market curve: generating, innovating, commercializing, commoditizing, attenuating. We also employed the sources of new capital development (NCD) as the functions of new business: marketplace capital development (MCD), organizational capital development (OCD), human capital development (HCD), information capital development (ICD), and mechanical capital development (mCD). As may be noted, the third dimension, or *"z dimension,"* is represented by the diagonal cells addressed: new science, new technology, new business, new services, new products.

Table 1-1. Concurrency—The MCD Possibilities Processing Paradigm (Initiative)

MARKET CURVE COMPONENTS

NCD FUNCTIONS	Generating	Innovating	Commercializing	Commoditizing	Attenuating
MCD	New Science				
OCD		New Technology			
HCD			New Business		
ICD				New Services	
mCD					New Products

With people representing each of the components and functions involved in marketplace positioning, or MCD, we processed simultaneously and interdependently. That is to say, people who processed interdependently with their operations had the opportunity to process interdependently with other people and other operations. As may be viewed, all operations processed simultaneously. Even before the new science cell had made a handoff to the new technology cell, the latter had begun to process. Similarly, before the new technology cell made a handoff to the new business cell, the latter was already processing. Of course, as these handoffs occurred, true interdependent processing ensued (see Table 1-2). All interactions were simultaneous and interdependent.

**Table 1-2. Concurrency—The MCD Possibilities
Processing Paradigm (Interdependent)**

MARKET CURVE COMPONENTS

NCD FUNCTIONS	Generating	Innovating	Commercializing	Commoditizing	Attenuating
MCD	New Science				
OCD		New Technology			
HCD			New Business		
ICD				New Services	
mCD					New Products

The differential implications of these two paradigms are profound. The sequential probabilities planning paradigm enables us to achieve our goals. We get exactly what we have planned for!

The concurrency possibilities processing paradigm enables us to achieve unlimited opportunities. We get far more than we processed for!

In the following pages, we will view the wondrous array of possibilities. For example, while the probabilities paradigm yields, at best, improvements on standard products, the possibilities paradigm yields a potentially infinite array of virtual products, services, solutions, partnerships, and constellations waiting to be tailored to customers' needs and requirements. Reduced, probabilities yields are infinitesimal. Elevated, possibilities yields are potentially infinite.

II

Possibilities Processing

Possibilities processing systems: inductive, deductive, generative, interdependent, and possibilities processing.

2 DNA′ and Inductive Intelligence

The science of possibilities addresses the limitations of the science of probabilities. This can perhaps be illustrated best by a review of the possibilities model, including possibilities conditions and standards. Our review will also serve as introduction to this chapter's discussion of DNA' and inductive intelligence.

THE POSSIBILITIES MODEL

The vital process-centric nature of possibilities science (and, indeed, all phenomena) informs the possibilities model, which is presented inductively in Figure 2-1. The difference between possibilities and probabilities is found in the quality of the dimensions and their interdependent relationships in continuously evolving processing systems. The operative words here are *"processing systems"*: all dimensions are defined as processing systems. Specifically, the culminating function is to free the changeability in phenomena. This requires the highest levels of phenomenal information, levels that enable us to relate and empower phenomena in order to release them. It also requires the highest levels of processing systems—interdependent phenomenal processing systems.

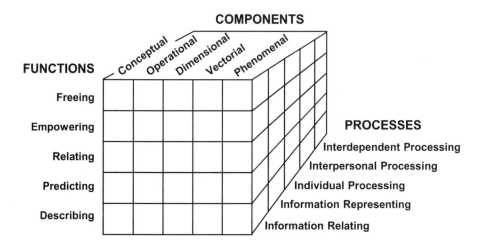

Figure 2-1. The Science of Possibilities (Inductive)

The possibilities model may also be presented deductively, as shown in Figure 2-2. Here the components are driven from the phenomenal level down: phenomenal, vectorial, dimensional, operational, conceptual. At the same time, the interdependent processes may be driven from the interdependent level down: interdependent, interpersonal, individual, informational. Deductive modeling is powerfully leveraged when phenomena are already known to us.

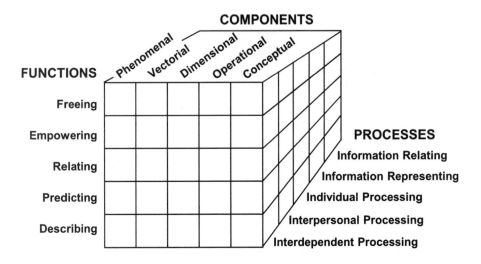

Figure 2-2. The Science of Possibilities (Deductive)

The possibilities conditions may be viewed in Figure 2-3. The conditions are the environments within which the phenomena are *nested*. As we can see, the conditions have their own unique operations: functions, components, and processes. If the conditions are freeing, then possibilities science can work its magic. Free-enterprise economics maximizes free choice for individual corporate entities. Free and direct democratic governance maximizes free choice for individual human phenomena. Free organizations maximize free choice in aligning resources for continuously evolving marketplace and governance objectives. Within the context of these conditions, the freeing functions of possibilities science may be served.

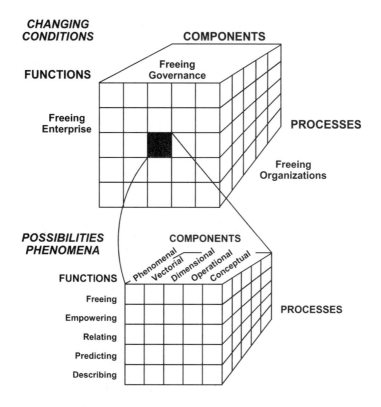

Figure 2-3. The Conditions of Possibilities Science

We can readily see possibilities conditions in the environment of our science of possibilities and the phenomena it generates: the freeing conditions of free-enterprise capitalism; the freeing conditions of direct-representation democracies; the freeing conditions of self-organizing, *"possibilities organizations."* These freeing conditions not only allow possibilities to be generated, but also enable them to be generated; moreover, in interdependent processing with the phenomena, the conditions themselves contribute to generating possibilities.

Possibilities standards may be viewed next in Figure 2-4. The standards are the measures of performance required to meet the freeing functions. The standards also have their own unique operations. Whereas the probabilities measures stress uniformity, the possibilities measures emphasize diversity. These measures maximize the range of performance. Ultimately, diversity is transformed into measures of changeability as the phenomena evolve in continuously changing forms. The standards are simply measuring what the new science is all about: possibilities.

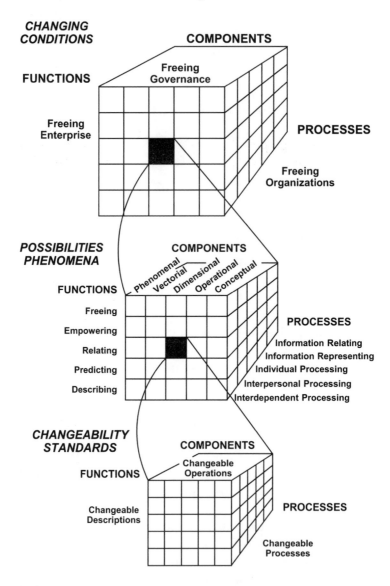

Figure 2-4. The Standards of Possibilities Science

Again, we can readily see diversity and changeability standards in the measures of operations for objectives: the diverse and changeable functions, components, and processes. These changeability standards not only allow possibilities to be measured, but also enable them to be generated; and, in interdependent processing with the phenomena, the standards themselves contribute to generating possibilities. All operations are processing systems: functions, components, processes, conditions, standards.

Source Cells and Hybrid Cells

We may view the possibilities science model by its *"source cells"*: the cells that cut diagonally across the model (see Figure 2-5). These cells are *"pure processing cells"* dedicated exclusively to the scientific functions:

- *Freeing functions* discharged by phenomenal information components enabled by I^5 interdependent processing;

- *Empowering functions* discharged by vectorial information components enabled by I^4 interpersonal processing;

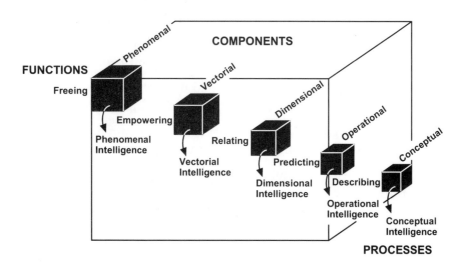

Figure 2-5. Source Cells of Possibilities Science

31

- *Relating functions* discharged by dimensional information components enabled by I^3 individual processing;

- *Predicting functions* discharged by operational information components enabled by I^2 information representing processes;

- *Describing functions* discharged by conceptual information components enabled by I^1 information relating processes.

These are illustrations of pure source cells. This means they are the prepotent combinations of information components and human processes dedicated to discharging their socio-genetic possibilities functions.

All other cells are hybrids of these pure source cells, and are composed of varying combinations of functions, components, and processes. A random sample is illustrated in Figure 2-6 and includes the following:

- Freeing functions discharged by conceptual information components enabled by I^1 information relating processes.

- Empowering functions discharged by dimensional information components enabled by I^3 individual processing.

- Relating functions discharged by phenomenal information components enabled by I^5 interdependent processing.

- Predicting functions discharged by vectorial information components enabled by I^4 interpersonal processing.

- Describing functions discharged by operational information components enabled by I^2 information representing.

Again, these are random illustrations of hybrid cells. This means that any combination of information-modeling components enabled by human processing systems may be dedicated to any possibilities functions. Ultimately, it means that any phenomena may be empowered and freed by possibilities science to seek their own changeable destinies.

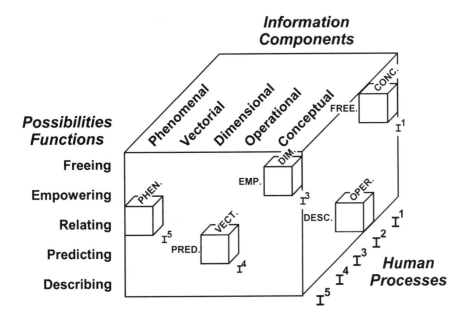

Figure 2-6. Possibilities Science Hybrid Cells

DNA´—DEVELOPMENTALLY NESTED ANALOG

If we look at the history of humankind, we find that human processing has been almost exclusively inductive. Certainly, such processing has been a limitation upon the rapid evolution and elevation of brainpower; however, this does not mean inductive intelligence cannot be useful. We may view in Figure 2-7 the essential origins of inductive intelligence, starting at the conceptual level. Ninety-nine percent of human processing has been conceptual. Most of the remaining one percent has been operational. Few have utilized the higher levels of dimensional, vectorial, and phenomenal intelligence in either an inductive or deductive way. In the pages that follow, inductive intelligence will be introduced in a manner that leads to the utilization of higher levels of intelligence.

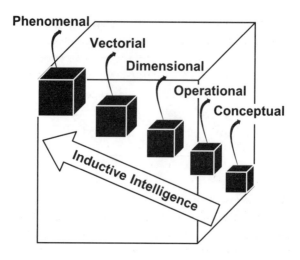

Figure 2-7. Inductive Intelligence

"DNA" is synonymous with "life." It is an abbreviated term for *deoxyribonucleic acid.* DNA is a spiral, double-stranded molecule found grouped in the nucleus of organic cells. It is the substance that "architects" the genetic information for growth. As such, this substance allows us to induce the entire organism, or phenomenon, from just a snippet of material. In this respect, DNA is our analog model for growth scaling and inductive intelligence (see Figure 2-8). Of course, we must also recognize a parallel and interactive

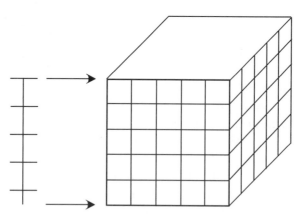

Figure 2-8. DNA and Growth

phenomenon known as *the phenomenological experience of science;* that is, the phenomenology of how the scientist experiences the phenomenon. From the scientist's observations, then, we may plot the movement of phenomena and scientist, within and between systems. This enables the growth of both the phenomena and the scientist. However, even this phenomenon of inductive reasoning may exist in a larger context of deductive reasoning. Indeed, it is this deductive intelligence that makes our inductive reasoning *intelligent.* To be sure, it is the reasoning of each—deductive and inductive—that empowers the intelligence in both.

In this context, we employ the term "DNA′" to represent our **"Developmentally Nested Analog"** system. Like all DNA, this DNA′ generates the genetic coding for growth. Simply stated, DNA′ is governed by one principle: **All processing systems are developmentally *nested* in other processing systems.** Deductively, the systems are *nested* in higher-order systems. However, depending upon our intentionality, they may be rotated to nest in lower-order systems. Inductively, the systems may be developed in any order or context we choose. The analog system simply relates the part to the whole—the strand of DNA to growth or life!

If we can scale, we can reason inductively by inducing generalizations. We may induce the whole from the part with external scaling (see Figure 2-9). Or we may induce the whole from the part

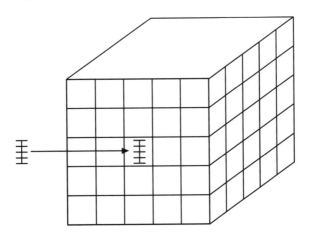

Figure 2-9. Inductive Intelligence

with internal scaling. In other words, we can induce models and relationships within and between models from DNA' scales.

For example, marketplace capital development (MCD) intelligence may be illustrated on the face of the marketplace positioning model, as shown in Figure 2-10. Here scales of the corporation's capabilities in technological terms (marketplace, organization, human, information, mechanical; or MOHIm) are compared with scales of the customer-driven marketplace requirements (MOHIm). In the illustration, an information technology company is currently positioned to meet its information-technology capabilities and the market's requirements for information capital development. The company projects its future marketplace positioning to meet information capabilities and organizational requirements. As may be noted, both current and future positioning are scaled inductively in a deductive marketplace environment.

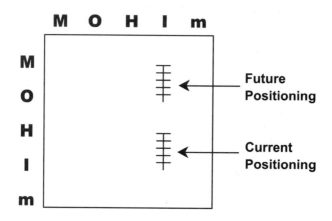

Figure 2-10. MCD Intelligence

In turn, organization capital development (OCD) intelligence is illustrated on the face of the OCD model, as shown in Figure 2-11. Here, organizational components (leadership, marketing, resources, technology, production; or LMRTP) are matrixed with marketplace functions (policy, executive, management, supervision, delivery; or PEMSD) inherited from MCD. As illustrated, in the currently supervised technology (T) component, the delivery (D) personnel aspire to the functions of self-supervised (S) performance. Again, we may note the type of scaling and its context: the supervisory phenomena are scaled inductively within a deductively derived OCD environment.

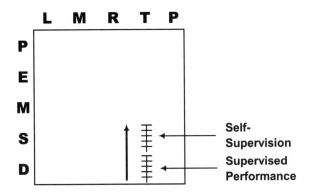

Figure 2-11. OCD Intelligence

Similarly, human capital development (HCD) intelligence is illustrated on the face of the HCD model (see Figure 2-12). Human components (goaling, inputting, processing, planning, outputting; or GIPPO) are matrixed with organization functions (LMRTP) inherited from OCD. In the illustration, a manager has aspired to "goal" (G) the resources (R) for the systems outputs (O). The systems phenomena are scaled inductively within a deductively derived HCD environment.

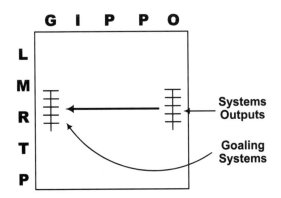

Figure 2-12. HCD Intelligence

The foregoing illustrations were presented in terms of relationships with phenomena. The growth scaling that bridges the gaps within and between phenomena is illustrated below in Figure 2-13. Notice that we may scale the growth program in precisely the same manner that we scale the phenomena. Indeed, the growth program is the relating phenomenon.

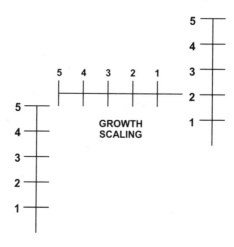

Figure 2-13. Intelligence and Internal Growth

The growth scaling that bridges the gap between phenomena is depicted in Figure 2-14. As we can see, the growth program enables growth between cells.

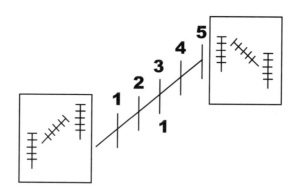

Figure 2-14. Intelligence and External Growth

Ultimately, we may relate the scales within any one of the cells of our systems with the scales of any or all of our systems (see Figure 2-15). Again, it is to be emphasized that all of these illustrations are scaled inductively within a deductively derived environment. The process of possibilities science is just that: **a process**. It emphasizes continuously evolving and spiraling sets of inductive generalizations and deductive discriminations related interactively by the results from testable hypotheses.

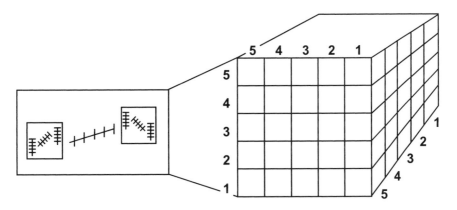

Figure 2-15. Inductive and Deductive Intelligence

Inductive Processing: A Case Study

Initially, possibilities scientists process inductively. This means that they move developmentally from descriptions or observations of a stable body of phenomena, through predictions or probabilities statements about the likelihood of the phenomena expressing themselves, to the hypothetical constructs or models that explain and control the phenomenal occurrence. These are the historical inductive functions of probabilities science: description, prediction, explanation and/or control.

Let us illustrate with the phenomena to which Dick Pierce and Alex Douds of Human Technology, Inc., related in the following situation. They were assigned to develop an empowerment program in I^5 interdependent processing skills. Beyond training, this meant conceptualizing all units and individuals in terms of HCD as defined by I^5.

These possibilities scientists proceeded inductively to develop the I^5 empowerment model. They illustrated inductive processing, or modeling, with clockwise rotations: old functions were rotated to become new components, and old components were rotated to new processes as new functions were introduced. Thus they arrived at the very notion of interdependent processing through inductive modeling.

Pierce and Douds began by making linear representations of information relating systems, or I^1, those systems that transform conceptual information into operational information. As illustrated in Table 2-1, the information-relating dimension was scaled in terms of its operations: functions, components, processes, conditions, standards.

Table 2-1. Information Relating Systems

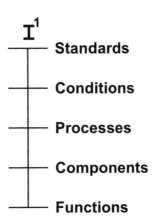

$$I^1$$

— Standards

— Conditions

— Processes

— Components

— Functions

The possibilities scientists continued inductive modeling by rotating inductively to make a matrix of information-relating components, or I^1, dedicated to information-representing functions, or I^2 systems, which transform operational information into dimensional information (see Table 2-2).

Table 2-2. Information Relating Matrix

I^2 FUNCTIONS	I^1 COMPONENTS				
	F	C	P	C	S
Phenomenal					
Vectorial					
Dimensional					
Operational					
Conceptual					

The information-representing dimension was scaled in terms of modeling systems: phenomenal, vectorial, dimensional, operational, conceptual. The objective of this matrix was defined as follows:

Information-representing functions are discharged by information-relating components.

In this way, the scientists defined phenomena operationally before they represented them multidimensionally.

The possibilities scientists continued inductive modeling by rotating to accomplish individual processing functions, or I^3: information-representing components, or I^2, dedicated to accomplishing I^3; information relating processes, or I^1, enabling the I^2 components (see Figure 2-16).

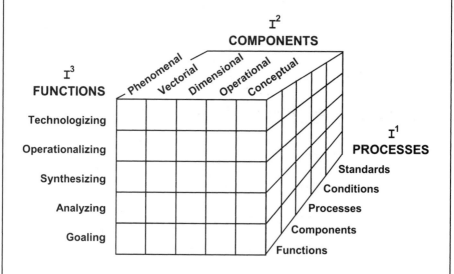

Figure 2-16. The Individual Processing Model

Next, the individual processing dimension was scaled in terms of its processes: goaling, analyzing, synthesizing, operationalizing, technologizing; or GASOT. This model of information representing was defined as follows:

Individual processing functions are discharged by information-representing components enabled by information relating processes.

In this way, the scientists represented phenomena multidimensionally before they processed them individually.

The possibilities scientists continued inductive modeling by rotating to accomplish interpersonal processing functions, or I^4: individual processing components, or I^3, dedicated to accomplishing I^4; information representing processes, or I^2, enabling the I^3 components (see Figure 2-17).

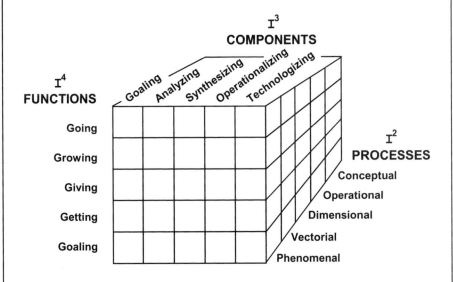

Figure 2-17. The Interpersonal Processing Model

Again, functions were scaled, with the interpersonal processing dimension scaled in terms of its processes: goaling, getting, giving, growing, going. This model of individual processing was defined as follows:

> *Interpersonal processing functions are discharged by individual processing components enabled by information representing processes.*

In this way, the scientists processed phenomena individually before they processed them interpersonally.

Now the possibilities scientists extended inductive modeling by rotating to accomplish interdependent processing functions, or I^5: interpersonal processing components, or I^4, dedicated to accomplishing I^5; individual processing systems, or I^3, enabling the I^4 components (see Figure 2-18).

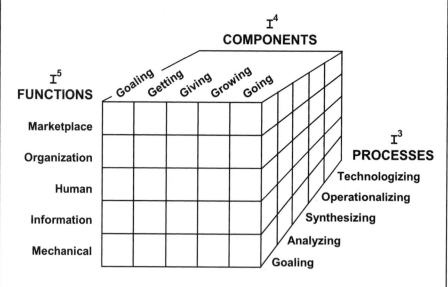

Figure 2-18. The Interdependent
Processing Model

Next, the interdependent processing dimension was scaled in terms of its processing systems: marketplace, organization, human, information, mechanical. This model may be defined as follows:

> *Interdependent processing functions are discharged by interpersonal processing components enabled by individual processing systems.*

In this way, the scientists processed interpersonally before they processed interdependently.

Finally, the possibilities scientists culminated inductive modeling by rotating to accomplish phenomenal processing functions: interdependent processing components, or I^5, dedicated to accomplishing phenomenal functions; interpersonal processing systems, or I^4, enabling the I^5 components (see Figure 2-19).

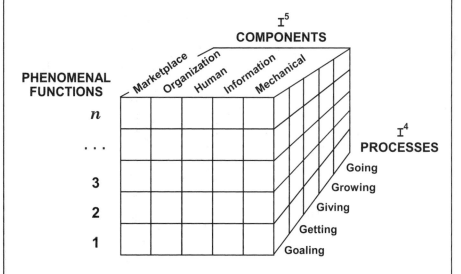

Figure 2-19. The Interdependent Phenomenal Processing Model

The phenomenal processing dimension was then scaled in terms of its operations: 1, 2, 3, ..., n. This model of interdependent phenomenal processing was defined as follows:

> *Phenomenal processing functions are discharged by interdependent processing components enabled by interpersonal processing systems.*

Thus the scientists processed interdependently in order to process phenomenally.

This is the critical level of interdependent processing systems. It is here that we generate images of all phenomenal possibilities—steps, tasks, stations, teams, units, organizations, marketplaces, and the like.

To sum, the possibilities scientists dedicated their interdependent processing systems to generating phenomenal possibilities (see Figure 2-20). This is the essence of inductive modeling.

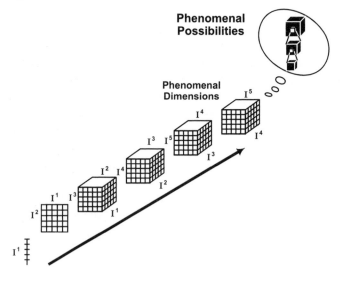

Figure 2-20. Interdependent Phenomenal Processing—Inductive Modeling

These interdependent phenomenal processing systems enable us to process within, between, and among all phenomenal systems.

Through processing inductively to generate new models of phenomena, the possibilities scientists learned to live in the phenomena they generated. They learned to process to innovate these new phenomena. They dedicated themselves to the proposition that all dimensions of all models are processing systems.

Inductive Processing: Applications

In our cultural capital development (CCD) project,* we did lots of growth scaling: scaling to enable people to grow; scaling to enable the growth of the organizations to which people were dedicated.

To survive, an organization must be a living, learning, processing entity. The historic mission of the organization was fairly straightforward—to become a *"productive agency"* for aligning resources. Consequently, the organization's role was to align its functions (or intentions), its resources, and its processes to implement marketplace positioning in the context of changing marketplace requirements. Today, however, most organizational units operate as if they have *"functional autonomy"*: they carry on with *"conditioned"* inherited missions! This will not work in the twenty-first century, which requires a growthful organization, one that continually realigns itself with continuously *"repositioned"* missions. Such an organization is, essentially, a *"thinking organization."*

---

* For more information on this CCD project, see Volume I of *The New Science of Possibilities.*

What follows are organizational alignment technologies or methodologies which we have employed in projects to tailor continuous alignment of the operations of continuously processing organizations. We begin by illustrating a system for aligning projects within or between organizations. The system's label, SALT,[*] is an acronym for inductive modeling: scaling, aligning, linking, transforming. SALT is a groupware application designed to align intra-enterprise and inter-enterprise resources to enhance organizational productivity.

Goaling by Scaling

We begin with goaling (see Table 2-3). Here we ask project leaders to employ a simple nominal (or naming) scale in order to describe the intended outputs of each project. Scaling gives us operational measures with which to guide the alignment of projects and evaluate their success. Project leaders can list the goals of their projects in nominal scales. These scales are a representation of the values of each project and, when viewed across projects, the values of the organization.

Table 2-3. Goaling by Scaling

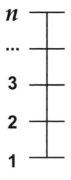

---

[*] Carkhuff Thinking Systems under the direction of Christopher J. Carkhuff and David Meyers.

Inputting by Scaling

Next, we may elicit input from all individuals or units affected by or affecting the intended project outputs (see Table 2-4). Again, we may employ simple nominal scales to describe measures of requirements for the intended outputs. These requirements tell us what resources will be needed to accomplish the desired outputs. Inputting by scaling results in the information we need in order to compare project values to project requirements. With this information, we can analyze the relationship resources and goals so we can make decisions about the alignment of precious resources. In other words, we now have the information necessary for maximizing values and minimizing resources across projects.

Table 2-4. Inputting by Scaling

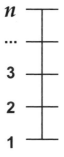

Processing by Aligning

We now align the scaled projects (see Figure 2-21). We may align these scaled projects by employing *"set theory"* to generate an inclusive scale and thus incorporate all individual project scales. As illustrated below, the *"across-project"* alignment scale is related orthogonally to the individual project scales. Processing by aligning scales enables us to see how our projects relate and how our organization is currently aligned.

It also enables us to see opportunities for the realignment of projects and the realignment of our organization to complete these projects.

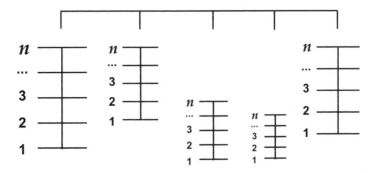

Figure 2-21. Processing by "Set-Theory" Alignment Scaling

Planning by Linking

Next, we plan by linking multiple project scales (see Figure 2-22). We link this information by connecting related project

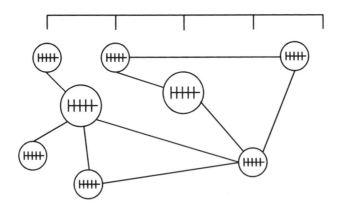

Figure 2-22. Planning by Linking

scales and the people responsible for these scales. Planning by linking will inform project managers of opportunities to transform the current organizational alignment into a more effective and efficient realigned organization, one with opportunities to realign project intentions, project resources, project processes, communication links, and communication flow.

Outputting by Transforming

Finally, we will make transformation decisions and implement transformation action programs. We will develop and implement programmatic courses of action to realign the project in our organization (see Figure 2-23). These changes will result in a realigned organization. When we recycle this project alignment process, we will see that our project scales have changed, reflecting the realignment of our organization: realigned intentions, realigned resources, realigned processes, realigned communication links, and realigned communication flow. Outputting by transforming in the final phase is a cycle of continuous alignment of projects across an organization.

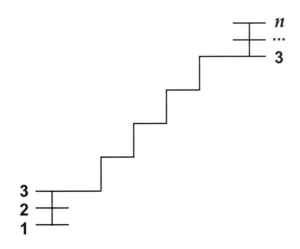

Figure 2-23. Outputting by Transforming

It is this organizational alignment processing system that we label "SALT." As mentioned earlier, SALT is an acronym for scale, align, link, transform. In this case, Project-Level SALT is an example of an OCD methodology for organizational realignment (see Figure 2-24).

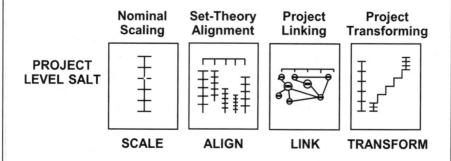

Figure 2-24. Project-Level Alignment by SALT

A workgroup application of SALT was designed by Carkhuff Thinking Systems (CTS) for a major software producer. The software company had a serious problem. It had created a market around its product by simultaneously creating partnerships for the product, and now the partners were getting most of the value for the product's servicing. Since the partnership program was sophisticated, the producer was prevented from competing directly with its partners in many areas where it could be successful. This meant that a portion of the producer's resources had to be positioned adequately upstream: from there, the producer could lead the market it had created.

The producer's Services Group took charge of developing the business around this new positioning, with the core products to be designed by CTS. Working with the services organization, CTS used its models to identify the most powerful and differentiated marketplace requirements. Since the partnership problem existed in almost every medium and large corporation, the key to the solution was identified: the Services Group had been disallowed from owning its own space.

Most of the producer's partners had the same organizational problems positioning themselves as customers who were expected in the producer's groupware. In other words, the partners became the producer's best customers. In this context, SALT was designed to be a qualitatively differentiated groupware application: it focused upon creating alignment throughout the supply chain into the customers' operations at increasingly more powerful levels of performance.

The alignment of company resources positioned the producer to actualize the value of team collaboration and networking. The Services Group positioned itself as the provider of a solutions package—software, education, methods— targeting enterprise and inter-enterprise alignment. Over time, it became clear to the executives and managers involved that they were actually innovators in a newly emerging class of market offerings.

In the process, we also designed SALT for higher-level organizational functions beyond aligning projects: team-level alignment, division-level alignment, company-level alignment, marketplace-level alignment. As shown in Figure 2-25, the SALT processing methodology becomes more complex as larger entities are more accurately represented by *multi-nested* scales and models. However, and most important, this methodology allows us to tailor alignment within, between, and among every level of organization. This provides us with a tool for continuous relationship engineering in organizations—the future of OCD.

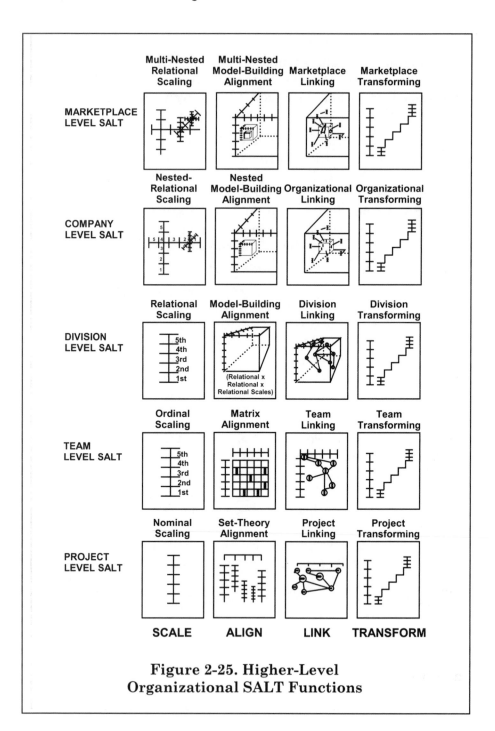

Figure 2-25. Higher-Level
Organizational SALT Functions

Some of the scales critical to business are illustrated in Table 2-5. As may be noted, these scales include the following: market optimization, extended enterprise relationship, business equation, process-centricity, knowledge-sharing. Again, the entire purpose of SALT processing is to maximize levels of functioning on all of these scales. At the highest levels, this means multiple companies, a constellation of businesses, an interdependent business equation, horizontal and vertical process integration, and inter-enterprise knowledge.

Table 2-5. Scales Critical to Business

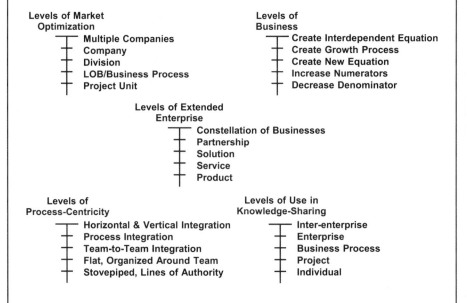

Levels of Market Optimization
- Multiple Companies
- Company
- Division
- LOB/Business Process
- Project Unit

Levels of Business
- Create Interdependent Equation
- Create Growth Process
- Create New Equation
- Increase Numerators
- Decrease Denominator

Levels of Extended Enterprise
- Constellation of Businesses
- Partnership
- Solution
- Service
- Product

Levels of Process-Centricity
- Horizontal & Vertical Integration
- Process Integration
- Team-to-Team Integration
- Flat, Organized Around Team
- Stovepiped, Lines of Authority

Levels of Use in Knowledge-Sharing
- Inter-enterprise
- Enterprise
- Business Process
- Project
- Individual

In short, SALT was designed to be a qualitatively differentiated application: it focused upon creating alignment across an enterprise; it extended throughout the supply chain into the customers' operations at increasingly more powerful levels of performance. The alignment of company resources is positioned to actualize the values of team collaboration and networking.

Based upon the SALT system, we also designed SALT Plus: an advanced modeling technology designed for developing a concurrent or parallel enterprise environment. Its main function is to align all resource dimensions (MOHIm) to a sophisticated process definition of the core business. The identification and process-oriented definition of the core business paradigm is the essential requirement for SALT Plus. This limits the candidates to companies that have a high degree of understanding of the core business and the interrelationships between the parts. For now, the early innovators are predominantly in the manufacturing industry where engineering and manufacturing functions are highly complex, and where the products are one line of business rather than multiple product-lines.

For example, the product life-cycles of complex products such as cars or planes can be defined as the master process against which the classification and alignment of the product development processes may be applied. This alignment applies to all populations that serve the master process: engineering, manufacturing, support, marketing. In time, a mature hierarchy of business populations and resource dimensions emerges.

SALT Plus was designed to model organizations where both the vertical and horizontal functions are in place and understood. That is to say, division and/or departmental functions are related to the life-cycle or supply chain of the master process. In this context, multidisciplined functional teams are positioned as well as supported by SALT Plus.

In this context, process improvement and process engineering can be performed by organizations, teams, or individuals with process-oriented objects: scaled matrices that inherit relationships within and between hierarchies. These process-oriented objects are designed to function along x, y, and z dimensions simultaneously. Objects inherit process interdependencies throughout the master process, between dimensions and populations as well as within their own local business areas.

IN TRANSITION

Interdependent processing prepares us to engage in continuous processing by relating our inductive and deductive processes interdependently and synergistically. The inductive approach is illustrated in the NCD example below (Figure 2-26). To view the figure inductively, we proceed from tasks to objectives, systems, and architecture, through to mission. In our next chapter, we will focus on *deductive* intelligence and its related approach.

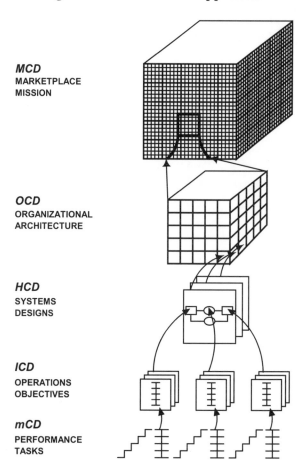

MCD
MARKETPLACE
MISSION

OCD
ORGANIZATIONAL
ARCHITECTURE

HCD
SYSTEMS
DESIGNS

ICD
OPERATIONS
OBJECTIVES

mCD
PERFORMANCE
TASKS

**Figure 2-26. Inductive-Intelligence
Illustration of NCD**

3 The Unifactoral Design and Deductive Intelligence

The operational dimensions of phenomenal processing are the universal operations of all processing: functions are discharged by components enabled by processes. Like the cellular operations of life, they are the cellular operations of ideation. The operations define the possibilities scientists just as the scientists define the operations. In other words, they define the conditions within which the possibilities scientists process.

The possibilities scientists process within phenomenal operations just as cellular operations process within the scientists. Ideationally, this means that possibilities scientists dedicate their processing potential to phenomenal functions. They become profoundly *"one"* with the phenomena.

The possibilities scientists cannot overemphasize their *"oneness"* with the phenomena: they live, breathe, nourish, and are nourished within, the phenomena. The inventor becomes the microchip; the physician, the virus; the manager, the organization; the worker, the task; the teacher, the learner; the child, the learning experience. All are possibilities scientists!

In inductive processing, the possibilities scientists generate new and powerful images of phenomena. In deductive processing, they innovate within the phenomena. This means that they relate to align with the phenomena; empower to enhance the potential of the phenomena; free the phenomena to seek their ever new and changing operations. These are the futuristic functions of possibilities science.

Deductive modeling innovates changes within, between, and among the phenomena. Thus, while inductive processing facilitates the evolution of phenomena, deductive processing accelerates *the revolution* of phenomena. In so doing, deductive processing and its derived hypothesis-testing is an accelerator of evolution. We no longer have to wait for the painfully slow increments of evolutionary change.

Possibilities science is the deductive source of all phenomenal possibilities. As such, it generates a unifactoral design for processing within which all processing systems are *nested* or *housed*. In this context, all phenomena may be related to, empowered, and freed to seek their own unique and changeable forms. The components that are transformed to discharge these functions are phenomenal information and its derivatives: vectorial, dimensional, operational, and conceptual information. Finally, the processes by which these information components are transformed into such functions are interdependent phenomenal processing systems and their phases: I^5 interdependent, I^4 interpersonal, I^3 individual, I^2 representational, I^1 interrelational.

Again, we may explore possibilities science as the source of all phenomenal possibilities by viewing its key cells, or *"source cells"* (see Figure 3-1). Source cells are *"pure"* cells that represent the interaction of similar or corresponding levels of dimensions: functions, components, processes. For example, in our highest, most powerful source cell, phenomenal processing is emphasized: phenomenal information components seek to free phenomena by way of interdependent phenomenal processing systems. Essentially, in this primary source cell, phenomenal components discharge freeing phenomenal functions through interdependent

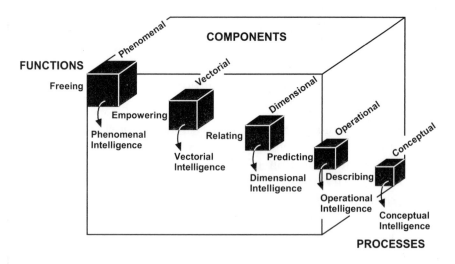

Figure 3-1. Source Cells of Possibilities Science

phenomenal processes. Accordingly, we may label this phenomenal processing system *"phenomenal intelligence."* Phenomenal intelligence is the source of deductive processing.

In a similar manner, other *source cells* reflect the interaction of corresponding levels of dimensions. The vectorial levels of information components are dedicated to empowering functions. They discharge these empowering functions through interpersonal processing with phenomena. We may label this vectorial processing system *"vectorial intelligence."* Vectorial intelligence extends our deductive processing.

Likewise, the dimensional levels of information components are dedicated to relating functions. They discharge these relating functions by way of individual processing. We may label this dimensional processing system *"dimensional intelligence."* Dimensional intelligence further extends our deductive processing.

The operational levels of information components are dedicated to predicting functions. They discharge these predicting functions by representing images of the phenomena. We may label this operational processing system *"operational intelligence."* Operational intelligence introduces discriminative learning.

Finally, the conceptual levels of information components are dedicated to describing functions. They discharge these describing functions by interrelating to merge with the phenomena. We may label this conceptual processing system *"conceptual intelligence."* Conceptual intelligence implements conditioned responding programming.

Again, these are pure *source cells*. We illustrate them for purposes of phenomenal meaning. In practice, any combination of dimensions may converge to define a given cell. For example, phenomenal intelligence may be illustrated as follows: phenomenal information components may be dedicated to phenomenal freeing functions by interdependent processing systems. At the other extreme, conceptual intelligence may be illustrated as follows: conceptual information components may be processed for any of the describing functions by information relating processes. Clearly, these different cells would be *"mapped in"* on different axes in the possibilities science model.

We may view the intelligence systems in operation in Figure 3-2. As may be noted, our phenomenal, vectorial, dimensional, operational, and conceptual intelligence systems flow deductively. In doing so, they direct us. Our deductive models tell us exactly what phenomena and what levels of phenomena to process.

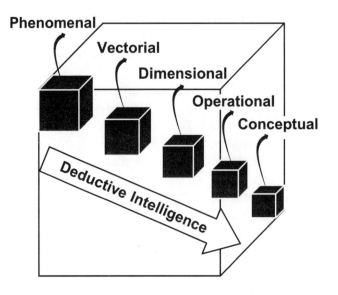

Figure 3-2. Deductive Intelligence

DEDUCTIVE MODELING

The highest level of interaction reflects the highest level of interdependency: phenomenal information components discharging phenomenal freeing functions through interdependent phenomenal processing systems, or I^5. We label this *"phenomenal intelligence"* because it defines and relates the universes of the phenomena involved; in order to free phenomena, it processes all phenomena to seek their own unique and changeable destinies. This initiates deductive processing.

At the highest levels, we can see phenomenal intelligence in operation in its transformation into phenomena. As may be noted, our source model, possibilities science, yields its first and highest-level *source cell*—the phenomenal intelligence cell. This cell positions phenomenal information components to accomplish phenomenal freeing functions by way of interdependent phenomenal processing systems.

We may view deductive intelligence in operation by modeling marketplace-driven vectorial systems. An overview of areas of new capital development (NCD) systems to be processed includes:

- Marketplace capital development, or MCD;
- Organizational capital development, or OCD;
- Human capital development, or HCD;
- Information capital development, or ICD;
- Mechanical capital development, or mCD.

In the illustrations that follow, the NCD system deductively models the alignment of all systems: MCD, OCD, HCD, ICD, mCD.

MCD

In the marketplace capital development model (Figure 3-3), the MCD functions, components, and processes are identified. The functions of MCD are derived from the market's requirements for NCD systems: MCD, OCD, HCD, ICD, and mCD. The marketplace

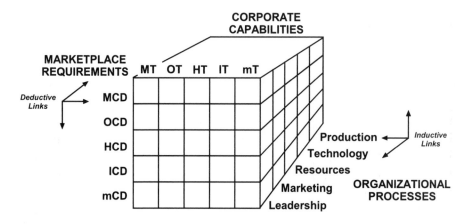

Figure 3-3. Marketplace Capital Development Model

of organizations is thus dedicated to fulfilling these NCD requirements. The MCD components are the corporate technologies available among organizations in the marketplace: MT, OT, HT, IT, and mT. These technologies are critical to fulfilling the requirements of the marketplace. The processes of the MCD model are organizational processing systems: leadership, marketing, resources, technology, production. To summarize:

> *Marketplace technologies are dedicated to new capital development enabled by organizational processing systems.*

The relationships of these marketplace requirements, technological capabilities, and organizational processing systems define MCD. The interaction of these dimensions defines MCD alignment. Once we realize the interrelational nature of MCD, we may link these dimensions with intentionality, doing so deductively, inductively, or functionally.

OCD

The functions of OCD are derived from the market's technology requirements (see Figure 3-4). These requirements are translated operationally into functional levels of the organization: policy, executive, management, supervision, and delivery. In other words,

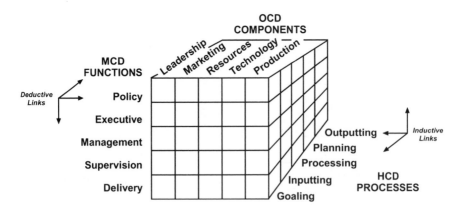

Figure 3-4. Organizational Capital Development Model

the resources of the organization will be dedicated to fulfilling these marketplace requirements. The OCD components are units of the organization and are derived from the processes of the MCD model: leadership, marketing, resources (and their integration), technology, and production. These organizational units are critical to fulfilling market requirements. The processes of the OCD model are introduced as HCD processes: goaling, inputting, processing, planning, and outputting. These HCD processes are essential for organizations to fulfill their goals. To summarize:

> *OCD components are dedicated to MCD functions*
> *enabled by HCD processing systems.*

The relationship of MCD functions, OCD components, and HCD processes defines OCD. The interaction of these dimensions defines OCD alignment. Once we realize the interrelational nature of OCD, we may link its dimensions with intentionality, doing so deductively, inductively, or functionally.

HCD

In the human capital development model (Figure 3-5), OCD components have been rotated to become the functions of HCD: leadership, marketing, resources, technology, production. Human

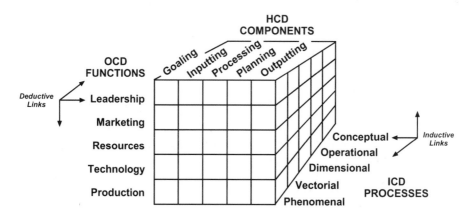

Figure 3-5. Human Capital Development Model

67

capital is thus dedicated to fulfilling these organizational goals. Similarly, the HCD processes of the OCD model have been rotated to become HCD components: goaling, inputting, processing, planning, and outputting. These human processing components are critical to fulfilling the goals of the organization. Finally, the information capital development (ICD) processes by which the HCD components discharge OCD functions are introduced: phenomenal, vectorial, dimensional, operational, conceptual. These ICD processes are essential for human processing. Note that again each lower-order ingredient is dedicated to enabling the achievement of a higher-order function:

> *HCD components are dedicated to OCD functions and enabled by ICD processes.*

These relationships of OCD functions, HCD components, and ICD processes define HCD. The interaction of these dimensions defines HCD alignment. Once we realize the interrelational nature of HCD we may link its dimensions with intentionality, doing so deductively, inductively, or functionally.

ICD

In the ICD model (Figure 3-6), HCD components have been rotated to become the functions of ICD. In other words, information capital is dedicated to servicing the requirements of thinking people. Likewise, the ICD processes of the HCD model have been rotated to become ICD components. These information components are critical ingredients in the service of human processing. Finally, mCD operations are introduced as enabling operationalizing processes: functions, components, processes, conditions, standards. These mechanical operations are essential to information capital processing. Again, note that lower-order ingredients are dedicated to achieving higher-order functions:

> *ICD components service HCD goals through mCD processes.*

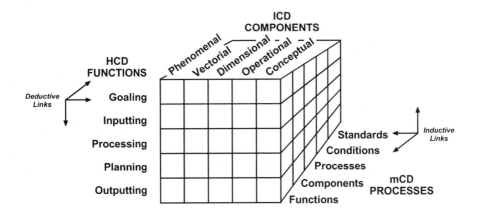

Figure 3-6. Information Capital Development Model

These interactions of HCD functions, ICD components, and mCD processes define ICD and its alignment. We may now link these dimensions with intentionality as well, deductively, inductively, or functionally.

mCD

The mechanical capital development model has ICD components that have been rotated to become the functions of mCD (see Figure 3-7). Mechanical capital, or mechanical tools, are dedicated to servicing information designs, or ICD. In turn, mCD processes of

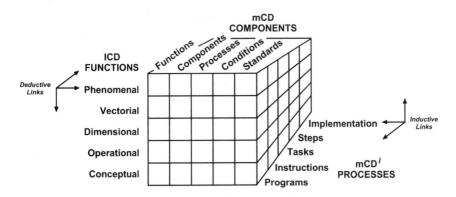

Figure 3-7. Mechanical Capital Development Model

the ICD model have been rotated to become mCD components. These mechanical components are critical to fulfilling information designs. Finally, new mCD′ programming processes are introduced: programs, instructions, tasks, steps, and implementation. These programmatic mechanical processes are essential to mechanical processing. Again, lower-order ingredients are dedicated to higher-order functions:

> *mCD components service ICD functions by mCD′ processes.*

The interaction of these dimensions and their relationships defines mCD and its alignment. We may now also link these dimensions with intentionality, doing so deductively, inductively, or functionally.

Deductive Processing: A Case Study

In practice, then, possibilities scientists apply interdependent processing systems deductively. This means that they move deductively from the phenomena which they or others have generated: they relate to align with these phenomena; they empower to enhance the potential of the phenomena; they free or release the potential of the phenomena. Again, these are the futuristic functions of possibilities science: relating, empowering, freeing.

Let us illustrate with the phenomena to which Chris Carkhuff and Don Benoit of Carkhuff Thinking Systems related in the following situation. They were assigned to employ I^5 interdependent processing systems to innovate breakthrough ideational models. Beyond generating models, this meant innovating within the models.

These possibilities scientists proceeded deductively to develop the I^5 innovation model. They illustrated deductive processing with phenomenal processing systems by rotating the dimensions counterclockwise: old components of higher-order systems became new functions of lower-order systems; old processes of higher-order systems became new components of lower-order systems; new processes were introduced. Carkhuff and Benoit began where inductive processing ended: by illustrating the phenomenal processing models generated by themselves or others (see Figure 3-8).

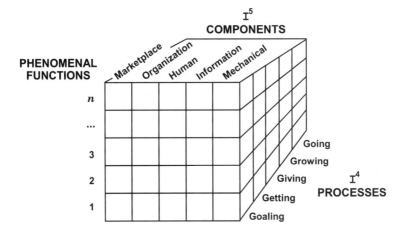

Figure 3-8. The Phenomenal Processing Model

At the highest level, then, possibilities scientists dedicate their I^5 interdependent processing systems to phenomenal possibilities that they or others have created:

> *Phenomenal possibilities are discharged by interdependent processing components enabled by interpersonal processing systems.*

This is the essence of phenomenal processing: the interdependent processing systems enabling us to process within the phenomenal systems. Accordingly, the possibilities scientists remain in this phenomenal processing mode unless they are unsuccessful in innovating within the phenomena, in which case they deduce the interdependent processing system to achieve higher levels of phenomenal processing.

At the next-highest level, interdependent processing functions (I^5) are discharged by interpersonal processing components (I^4) enabled by individual processing systems (I^3). We model such processing as shown in Figure 3-9.

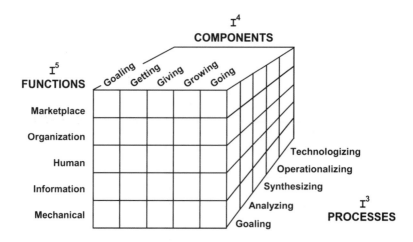

Figure 3-9. The Interdependent Processing Model

Here, the possibilities scientists dedicate their I^4 interpersonal processing systems to I^5 interdependent processing:

Interdependent processing functions are discharged by interpersonal processing components enabled by individual processing systems.

This is the essence of interdependent processing: the interpersonal processing systems empowering us to process within interdependent processing systems. Accordingly, the possibilities scientists engage in interpersonal processing in order to expand and elevate their interdependent processing systems.

At the next level, interpersonal processing functions (I^4) are discharged by individual processing components (I^3) enabled by informational representing processes (I^2). We model such processing as illustrated in Figure 3-10.

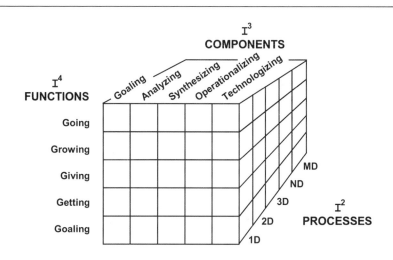

Figure 3-10. The Interpersonal Processing Model

Here, the possibilities scientists dedicate their I^3 individual processing systems to I^4 interpersonal processing:

> *Interpersonal processing functions are discharged by individual processing components enabled by information representing processes.*

This is the essence of interpersonal processing: the individual processing systems empowering us to process within interpersonal processing systems. Accordingly, the possibilities scientists generate higher-quality responses in order to achieve elevated levels of interpersonal processing.

At the following level, individual processing functions (I^3) are discharged by information-representing components (I^2) enabled by informational relating processes (I^1). This processing is modeled as shown in Figure 3-11.

The possibilities scientists now dedicate their I^2 information representing systems to I^3 individual processing:

> *Individual processing functions are discharged by information-representing components enabled by information relating processes.*

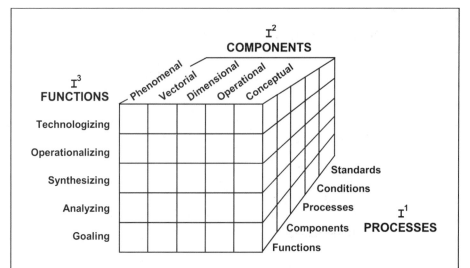

Figure 3-11. The Individual Processing Model

This is the essence of individual processing: the information representing systems empowering us to model information within individual processing systems. Accordingly, the possibilities scientists build higher-quality models in order to achieve elevated levels of individual processing.

At the next level, information-representing functions (I^2) are discharged by information-relating components (I^1). Table 3-1 presents an illustration of this level.

Here, the possibilities scientists dedicate their I^1 information relating systems to I^2 information representing systems:

Information-representing functions are discharged by information-relating components.

This is the essence of information representing: the information relating systems empowering us to define operational information within information representing systems. Accordingly, the possibilities scientists define higher-quality operations in order to achieve elevated levels of information representing.

Table 3-1. Information Relating Matrix

I^1 COMPONENTS

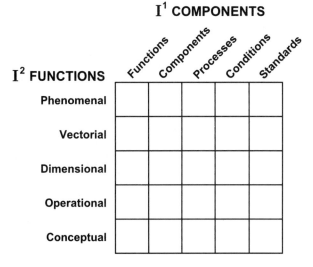

I^2 FUNCTIONS	Functions	Components	Processes	Conditions	Standards
Phenomenal					
Vectorial					
Dimensional					
Operational					
Conceptual					

Finally, the operations of information relating are scaled (see Table 3-2). As may be noted, the I^1 dimension is scaled in terms of its operations: functions, components, processes, conditions, standards.

Table 3-2. Information Relating Scale

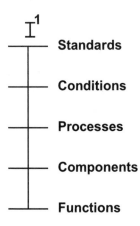

I^1
— Standards
— Conditions
— Processes
— Components
— Functions

To sum, the possibilities scientists dedicate their interdependent processing systems to innovating within the phenomenal possibilities that they or others have created. This is the essence of interdependent phenomenal processing—deductive modeling (see Figure 3-12). These interdependent processing systems enable us to process within, between, and among all phenomenal systems.

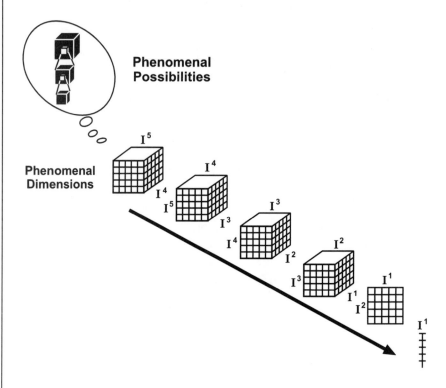

Figure 3-12. Interdependent Phenomenal Processing—
Deductive Modeling

Again, possibilities scientists cannot stress enough the power of interdependent deductive processing. Figure 3-13 represents the generating power of interdependent processing within phenomena. As we can see, the I^5 system merges with the phenomenal system and is dedicated to the phenomenal dimensions. Enabled by interpersonal processing, then, the possibilities scientists merge with the phenomenon and process its dimensions. In other words, they engage in interdependent phenomenal processing:

Continuous and partnered phenomenal processing for mutual benefit.

Indeed, both the phenomenon and its processing partners grow through this continuous interdependent processing.

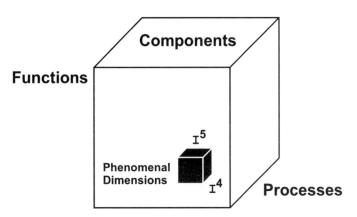

Figure 3-13. Interdependent Phenomenal Processing

Deductive Processing: Applications

In this section, we will review some of the innovations that Carkhuff Thinking Systems has produced through deductive processing. These center on the development of the NCD Generating Engine and a related application called Virtual Organization Reality.

THE NCD GENERATING ENGINE

Dedicated to new capital development, this software-generating engine is a key innovation in the development of organizational processing systems. Based upon scientific breakthroughs in intelligent modeling, the NCD Generating Engine integrates all processing systems—marketplace, organization, human, information, and mechanical—into an organizational processing system (see Figure 3-14).[*] Functionally, the Engine's operations replicate and integrate all of the operations of the NCD systems. They achieve the NCD functions (MCD, OCD, HCD, ICD, mCD) and dedicate to achieving these functions the NCD capacities in technologies (MT, OT, HT, IT, mT). The operations also enable the NCD capacities to achieve the NCD functions by enabling NCD processing systems:

- S-MP-R marketplace processing systems,
- S-OP-R organizational processing systems,
- S-P-R human processing systems,
- S-O-R information-modeling systems,
- S-R mechanical tooling systems.

With the NCD Generating Engine, we may process deductively, inductively, or idiosyncratically to tailor systems responses to meet any requirement. Representing and relating

---

[*] Carkhuff, C. J., Carkhuff, R. R., and Meyers, D. *The OCD Generating Engine.* McLean, VA: Carkhuff Thinking Systems, 1994.

all dimensions that currently make any organization productive and growthful, the Engine also fluidly incorporates all new or evolving ingredients that may accelerate the growth and development of an organization.

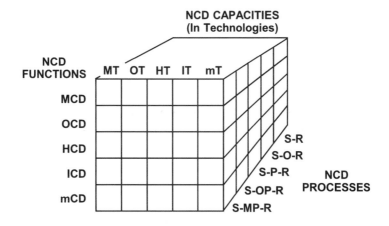

Figure 3-14. The NCD Generating Engine

At the most basic level, the NCD Generating Engine is an *"intelligent router"* of organizational information. At the most powerful level, it has the potential of processing over 30 billion interdependent organizational initiatives. We may think of the Engine's design as a hybrid of the best qualities of CASE, AI, Expert Systems, Neural Nets, OOPS, and Parallel Processing technologies. These technologies, however, do not suffice to describe the power of this new scientific platform.

The NCD Generating Engine is best understood using an organic analogy. Working with an organization's genotype, or genetic constitution, the Engine creates and then manipulates Organizational DNA, or ODNA. Its information output is Organizational RNA, or ORNA, which gives life to the organization. As a result of such capabilities, the Engine allows us to test any scenario. For example, we may rotate our vectorial systems internally as well as externally.

VIRTUAL ORGANIZATION REALITY

Perhaps the greatest transformation of the NCD Generating Engine has been *"Virtual Organization Reality,"* or *VOR:* a dynamic, graphically intuitive representation of organization realities adaptable to any environment or platform. Imagine, anyone may enter the virtual reality of this experience and view any cell, test any scenario, track any initiative. Most important, anyone can generate totally new and powerful organizations tailored to respond to spiraling changes in the marketplace. In essence, we can process interdependently to design continuously changing organizations.

At Carkhuff Thinking Systems, we have developed the breakthrough technologies for VOR.[*] This is a major step toward *Virtual Reality Networks*. VOR enables us to enter the full experience of any organization.

In the VOR system, we may experience flying into a specific space in an organizational enterprise:

- City space representing the marketplace environment;

- City blocks representing the inter-enterprise environment;

- Buildings representing the organizational enterprise;

- Floors of buildings representing units or teams of people;

- Rooms within floors representing individuals and groups of people.

In addition, we may see within and between the rooms processing objects and sub-objects that represent, in various shades of color, the NCD processing systems:

---

[*] Carkhuff, C. J., and Carkhuff, R. R. *Virtual Organization Reality.* McLean, VA: Carkhuff Thinking Systems, 1997.

- MCD, or marketplace-positioning objects;
- OCD, or organizational-alignment objects;
- HCD, or human-processing objects;
- ICD, or information-modeling objects;
- mCD, or mechanical-tooling objects.

In this manner, we can track firsthand the iterations of policy decisions, executive architecture, management systems, supervisory objectives, and performance tasks.

Based upon our NCD Generating Engine, then, the VOR application generates simple and intuitive multidimensional images and text in plain English. Furthermore, the VOR application *"clothes"* the phenotype with new qualitative forms of information as well as quantitative data from subsidiary databases. In short, the NCD Generating Engine gives birth to life in VOR (see Figure 3-15).

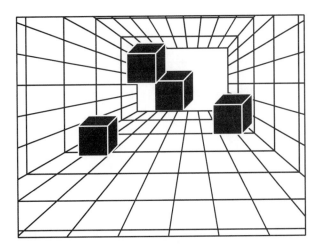

Figure 3-15. Virtual Organization Reality

In VOR, the user's view into the organization's phenotype, or observable constitution, depends upon the application chosen. For example, in a reengineering application, the user may view a graphic representation of people in a room, rooms on a floor, floors in a building, buildings on a block, or blocks in a city; a typical VOR room is shown in Figure 3-16. All of these views are derived from the organizational DNA and tailored to the user's unique perspective like the cells and organs of a living system.

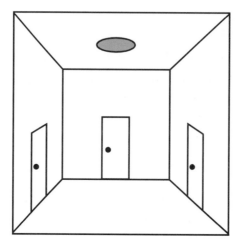

Figure 3-16. VOR Rooms

This tailoring allows VOR to simplify reengineering (as well as other processes), while its flexibility allows quick hypothesis-testing. For instance, VOR presents business processing cells as rooms of people, data, and things, with buildings representing departmental or divisional productivity models. The user may move the cells to fit desired goals, and enter and view the resulting solution. This enables multiple hypothesis-testing in a matter of minutes! And if the present options do not please, the user can simply change the values of the building, and instantly see the results as a repositioning of the rooms. Thus it is possible to define an ideal productivity model that actualizes inter- and intra-divisional operations *all in an afternoon!*

To sum, VOR provides an electronic processing environment that delivers concurrency to business constellations. This processing environment empowers users to activate their business dynamically with inherently interdependent process-centricity. Relatedly, VOR reduces the investment in planning time and resources significantly by automatically replacing planning models with processing models—models that can be adjusted to the changing environment. Consequently, the processing environment frees initiatives to optimize their value at any window of opportunity.

The electronic processing code is an information capital development system that creates its own graphic definition of itself. This DNA of universal processing functions interdependently and concurrently to fulfill organizational missions: connectivity, communication, collaboration, coordination, concurrency. This processing DNA also simplifies the complexity of all dimensions: because it has multidimensional knowledge of itself, computational formulas move in multidimensional space, reducing instruction sets to a minimum.

In this manner, VOR gives us a virtual experience of our organizational functions. Not only does it enable us to locate and communicate with various entities within the organizational constellation, it also allows us to relate to them within the framework of a common organizational processing language. Moreover, it brings to life all entities within the organizational constellation, allowing us to engage intimately and simultaneously and interdependently in its organizational processing systems.

Just imagine! The employee enters the VOR experience to model any scenario and to understand its implications. Thus, VOR not only brings the organization to life, but also brings the employees to "lifefulness": it rededicates them to the tasks-at-hand with an understanding of legacy learnings and future processing initiatives.

Internal Alignment

We may utilize VOR in strategic processing for organizational alignment with our marketplace positioning. For example, just as the vectorial systems are rotated externally, so may they be rotated internally. The illustrations that follow are straight rotations of the driving components and functions of internal delivery systems.

The MCD delivery systems are shown in Figure 3-17. As may be noted, the policy component drives the MCD (marketplace positioning) function; the executive component drives the OCD (organizational architecture) function; the management component drives the HCD (human systems) function; the supervisory component drives the ICD (information modeling) functions; the delivery component drives the mCD (mechanical systems) functions.

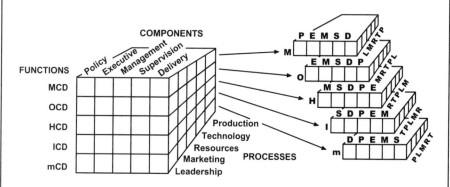

Figure 3-17. MCD Delivery Systems

Similarly, with the OCD delivery systems (see Figure 3-18), the leadership component drives policymaking functions; marketing drives executive functions; resource integration drives management functions; technology development drives supervisory functions; and production drives delivery functions.

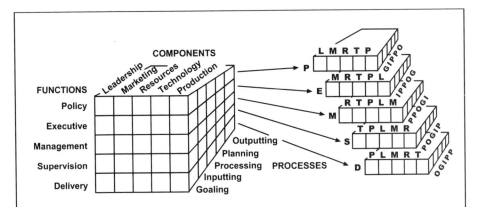

Figure 3-18. OCD Delivery Systems

In turn, the internal rotations with the HCD delivery systems yield the following component differentiation (see Figure 3-19): the goaling component drives mission-building functions; the analyzing component drives organizational architecture; the synthesizing component drives systems functions; the operationalizing component drives objective functions; and the technologizing component drives task functions.

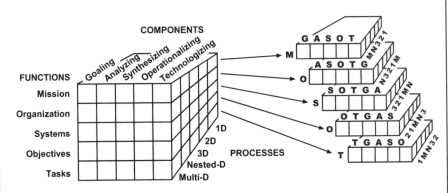

Figure 3-19. HCD Delivery Systems

Also, the internal rotations of the ICD delivery systems yield the following differentiated components (see Figure 3-20): multidimensional modeling drives goaling functions; *nested* modeling drives analyzing functions; three-dimensional modeling drives synthesizing functions; two-dimensional matrixes drive operationalizing functions; and linear scaling drives technologizing functions.

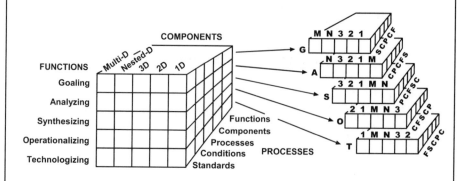

Figure 3-20. ICD Delivery Systems

Finally, the internal rotations of the mCD delivery systems find mechanical processing or conditioned responding driving the dimensionalizing operations (see Figure 3-21): standards drive multidimensional modeling functions; conditions drive *nested* modeling functions; processes drive three-dimensional modeling functions; components drive two-dimensional matricing; functions drive one-dimensional scaling.

Although the process is generated deductively, both the processes and the products of the processing may be highly variable; indeed, they may be diverse and changeable. In this context, let us assume that we have different images for deductively generated organizational processing systems.

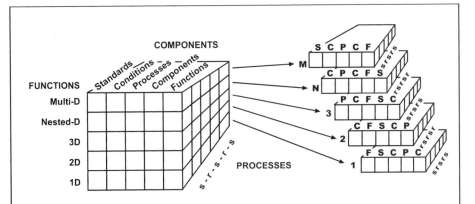

Figure 3-21. mCD Delivery Systems

For example, we may process interdependently within the leadership cell alone: the policy-positioning function discharged by the leadership component empowered by goaling processes (see Figure 3-22).

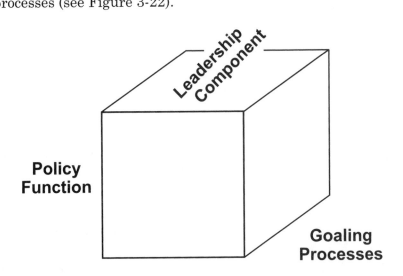

**Figure 3-22. Processing Interdependently
Within the Leadership Cell**

We may also develop an expanded vision of leadership by processing interdependently within the organizational components: leadership, marketing, resources, technology, production (see Figure 3-23). In addition, the human processes may be defined comprehensively: goaling, inputting, processing, planning, outputting. As may be noted, then, policymaking functions are now discharged by all organizational components enabled by comprehensive human processing.

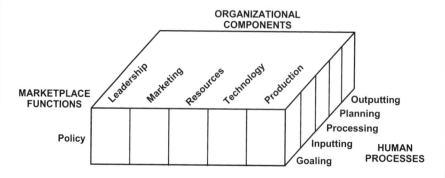

**Figure 3-23. Processing Interdependently
Between Organizational Components**

Further, we can align marketplace functions with leadership-driven components and functions to form a positioning team (Figure 3-24). With continuous relating to the marketplace, this team has the function of processing interdependently to transform relating functions into marketplace positioning. We may define the positioning team's phenomenal objective as follows:

> *Relating-positioning functions are discharged by leadership-driven components enabled by comprehensive human processing.*

We thus may label this positioning team *"The Possibilities Positioning Team."*

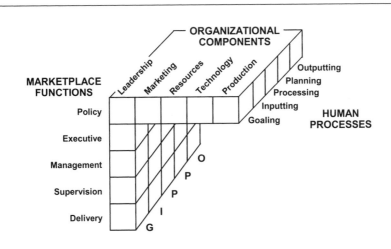

Figure 3-24. Possibilities Positioning Team

Next, the processing team may generate and align process-ing teams with this relating-positioning team (see Figure 3-25). Here we may note that the teaming systems are represented according to their differential contributions to accomplish the relating positioning:

- *Organizational alignment functions are discharged by market-driven components enabled by interdependent processing teams.*

- *Tailored customer-design and -solution functions are discharged by resource-integration components enabled by interdependent processing teams.*

- *Customized product and service functions are dis-charged by technology-driven components enabled by interdependent processing teams.*

- *Standardized product functions are discharged by production components enabled by interdependent proc-essing teams.*

We thus may label these organizational teams *"Possibilities Processing Teams."*

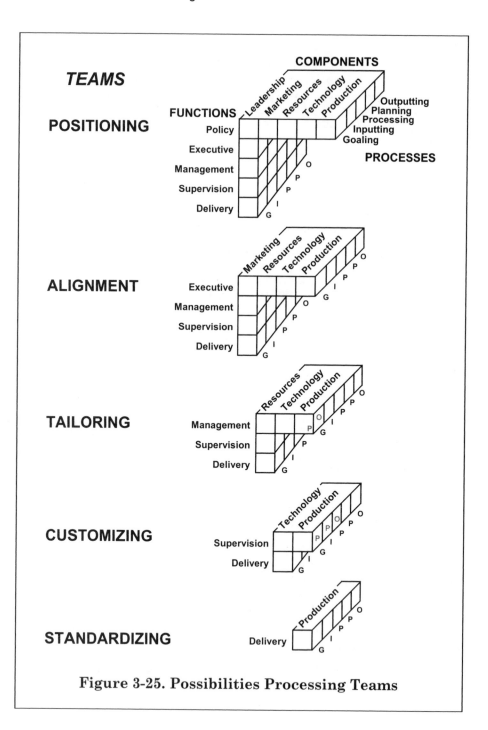

Figure 3-25. Possibilities Processing Teams

Within, between, and among any of these processing teams or units, then, are applied our I^5 interdependent processing systems to generate continuously changing phenomenal possibilities:

- Processing to generate continuous marketplace positioning;

- Processing to generate continuous organizational alignment;

- Processing to generate continuous customer tailoring;

- Processing to generate continuous product customizing;

- Processing to generate continuous product standardizing.

These organizational processing teams themselves work together synergistically. For example, as products are standardized, they enable the organization to reposition itself in the marketplace and, thus, to recycle the organizational processing teams.

IN TRANSITION

This chapter has presented many illustrations of deductive intelligence. It is important to remember that, in practice, our systems can be rotated inductively as well as deductively. Indeed, our systems can be rotated idiosyncratically in any direction to discharge any function. The great contribution of deductive intelligence is that it represents the processing of all entities. And it rotates—externally and internally—to direct any function! In short, it makes us intelligent by enabling us to process interdependently with all contributing entities.

We may view deductive modeling in the NCD application shown in Figure 3-26. To view the figure deductively, we proceed from mission to architecture, design, and objectives, through to tasks. This deductive-intelligence illustration may be compared to the inductive-intelligence example that concludes Chapter 2. Clearly, deductive intelligence is more powerful and economical in generating and testing hypotheses and modifying models.

The great contribution of deductive intelligence is its acceleration of evolution. Instead of inductively gathering data to populate our inductive generalizations, we may employ deductive intelligence to generate a plethora of immediately testable hypotheses. The results of these tests guide our future applications and tests in increasingly rapid cycles. Thus, we make things happen in a very brief period of time, things that may have taken centuries or even millennia to happen inductively. In this respect, deductive modeling creates time.

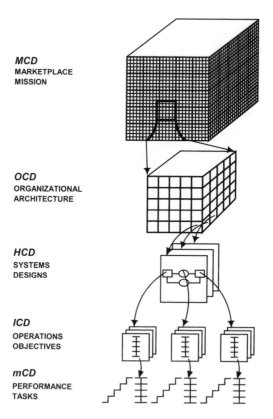

MCD
MARKETPLACE
MISSION

OCD
ORGANIZATIONAL
ARCHITECTURE

HCD
SYSTEMS
DESIGNS

ICD
OPERATIONS
OBJECTIVES

mCD
PERFORMANCE
TASKS

**Figure 3-26. Deductive-Intelligence
Illustration of NCD**

4 Interdependency and Generative Intelligence

Possibilities scientists are both generative and innovative processors. They process inductively to generate new and powerful models of phenomena. They process deductively to innovate these models of phenomena.

The possibilities scientists view the inductive and deductive processing systems in operation, as illustrated in Figure 4-1. We may note that the inductive processing systems culminate in generating new images of phenomena; in turn, the deductive processing systems culminate in innovating new images within, between, and among these images.

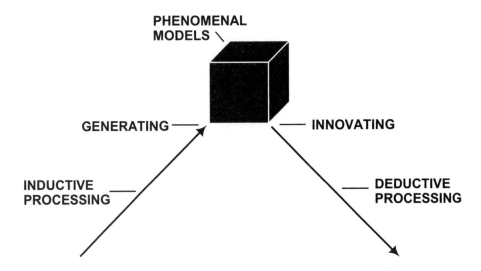

Figure 4-1. Inductive and Deductive Processing

The interactions of these inductive and deductive processes are the essence of possibilities science (see Figure 4-2). The results of each hypotheses-testing reflect back to innovate or modify the image of the phenomena: each test reflects a more accurate image of the phenomena; all tests move the phenomenal images toward convergence with the phenomena themselves. Thus, the possibilities scientists are continuously testing deductively to innovate prime phenomenal images.

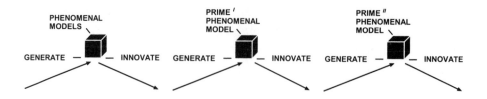

Figure 4-2. The Continuous Interactions of Inductive and Deductive Processing

For the possibilities scientist, the inductive and deductive processes are clear (see Figure 4-3). The inductive processes *generate* new responses through interdependent processing systems: I^1, I^2, I^3, I^4, I^5. By contrast, the deductive processes *innovate* new responses through interdependent processing systems: I^5, I^4, I^3, I^2, I^1. While the inductive processes generate responses incrementally, the deductive processes innovate responses decrementally: they turn to lower-order systems only upon realizing the limitations of higher-order systems, and they deliver lower-order systems to test hypotheses in support of building responses in higher-order systems.

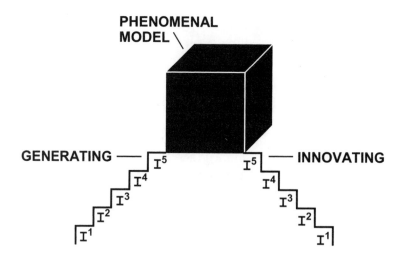

Figure 4-3. Inductive and Deductive Processing

In their full complexity, the possibilities scientists are processing inductively with three qualitatively different human processing systems (see Figure 4-4):

- *Stimulus-response (S-R) conditioned responding systems,* which define the operations of the conditioned responses;

- *Stimulus-organism-response (S-O-R) discriminative learning systems,* which represent or model dimensional images to replace the operational images of phenomena;

- *Stimulus-processor-response (S-P-R) generative processing systems,* which generate increasingly more powerful images of phenomena.

At the highest levels of the S-P-R systems, the inductive processors are processing with different phenomenal processing systems, such as the organizational (S-OP-R) processing systems. These phenomenal processing systems enable them to generate the phenomenal models.

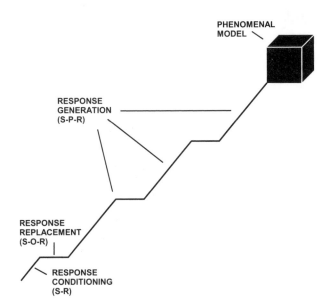

Figure 4-4. The Processing Systems of Inductive Processing

Deductively, the possibilities scientists are also processing with these qualitatively different human processing systems (see Figure 4-5):

- *S-P-R generative processing systems,* which empower the scientists to innovate within the phenomenal models;

- *S-O-R discriminative learning systems,* which enable them to innovate response-replacement images of the phenomenal models;

- *S-R conditioned responding systems,* which enable them to innovate new conditioned response operations.

Again, the S-P-R systems process interdependently with phenomenal processing systems to generate improved images of the phenomenal models.

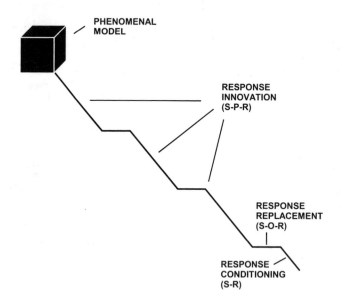

Figure 4-5. The Processing Systems of Deductive Processing

Inclusively, the possibilities scientists process interactively with all human processing systems (see Figure 4-6). This involves the following:

- Inductively building models by building responses;

- Deductively testing hypotheses for response performance;

- Interactively improving the images of the phenomenal models.

Interdependent phenomenal processing is a continuous and life-long process: continuous because the phenomena are constantly changing; life-long because the changes extend eternally to the lives of the phenomena and the processors.

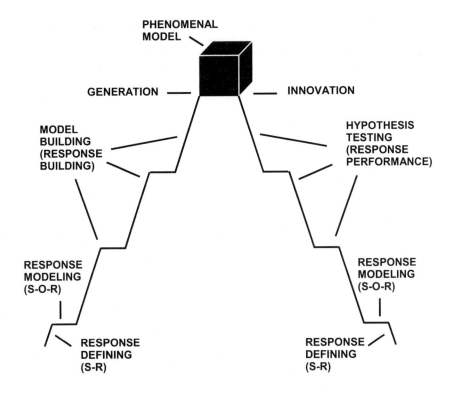

Figure 4-6. The Model Building and Hypothesis-Testing of Inductive and Deductive Processing

The interdependent phenomenal processing systems enable the possibilities scientists to process continuously with constantly evolving phenomena:

- *Relating* to align with the phenomena;

- *Empowering* to enhance the phenomenal potential;

- *Freeing* the phenomena to actualize their own changeable potential.

These are the functions of possibilities science. They enable the possibilities scientists to continuously generate and innovate phenomenal models. They enable the possibilities scientists to accelerate the movement of science and civilization.

Both deductive and inductive intelligence converge upon dimensional intelligence (see Figure 4-7). Basically, phenomenal and vectorial intelligence impose "directionality" upon dimensional intelligence. In turn, conceptual and operational intelligence build developmentally to dimensional intelligence. We may label this interactive circumstance *"generative intelligence"*: all vectors and factors converge upon these cells.

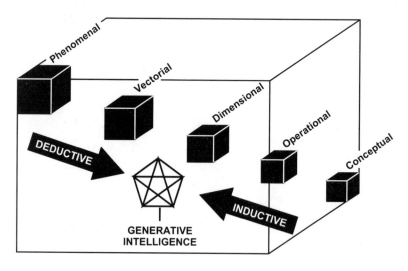

Figure 4-7. Generative Intelligence

Dimensional intelligence is the prerequisite for generative intelligence. With dimensional intelligence—whether deductively derived or developmentally induced—generative processing is possible. Without dimensional intelligence, generativity is not possible. With both deductively derived and developmentally induced dimensional intelligence, the highest levels of generative processing are possible.

Dimensional intelligence simply means that we have defined the phenomena—any and all phenomena that concern us—multi-dimensionally; that we have defined their functions, components, processes conditions, standards. Dimensional intelligence allows us to relate all of these dimensions interdependently. Moreover, dimensional intelligence empowers us to generate new and more powerful dimensions and, thus, new and more powerful phenomena. The generation of these new and more powerful dimensions is the core of generative processing.

GENERATIVE MODELING SYSTEMS

Generative processing systems are, quite simply, the interdependent processing systems that empower us to generate or create new images of phenomena. As shown in Figure 4-8, they include the following:

- I^1—*Information relating systems,* which enable us to define the operations of phenomena;

- I^2—*Information representing systems,* which enable us to model images of the phenomena;

- I^3—*Individual processing systems,* which enable us to process the phenomena individually to generate new images of the phenomena;

- I^4—*Interpersonal processing systems,* which enable us to process the phenomena interpersonally to generate more powerful images of the phenomena;

- I^5—*Interdependent processing systems,* which enable us to process the phenomena interdependently to generate the most powerful images of the phenomena.

While interdependently related, the generative processing systems allow us to enter at any phase of processing.

101

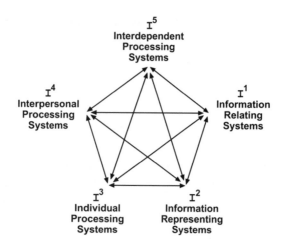

Figure 4-8. Generative Processing Systems

We begin developmentally with the information relating systems, which empower us to transform conceptual information about phenomena into operational information defining phenomena. These information relating systems are discharged by more basic interrelating systems that enable us to relate to all phenomena. Operationally, this may be expressed as follows:

> *Information-relating functions are discharged by interrelating systems that enable us to define phenomenal operations.*

In turn, the information representing systems empower us to transform this operational information into dimensional information or phenomenal representation:

> *Information-representing functions are discharged by information-relating components.*

Similarly, the individual processing systems build upon the information representing systems by incorporating the models they have produced in order to generate new images of the phenomena:

> *Individual processing functions are discharged by information-representing components enabled by information relating processes.*

Likewise, interpersonal processing systems build upon individual processing systems by incorporating individually generated images of the phenomena in order to generate more powerful images of the phenomena:

> *Interpersonal processing functions are discharged by individual processing components enabled by information representing processes.*

Finally, interdependent processing systems build upon interpersonal processing systems by incorporating interpersonally generated images of the phenomena in order to generate the most powerful images of the phenomena:

> *Interdependent processing functions are discharged by interpersonal processing components enabled by individual processing systems.*

Keep in mind that, even as they rotate dimensions, all I^5 systems are processing systems.

Again, the generative processing system may be entered in any phase. Each phase relates interdependently with all other phases. All phases relate interdependently with the phenomena they are processing.

Phenomenal Processing Systems

Phenomenal systems are the processing systems of the phenomena; to process phenomenal systems, we employ generative processing systems. As shown in Figure 4-9, the phenomenal processing·systems include the following:

- *Phenomenal systems,* which most closely approximate the phenomena themselves;

- *Vectorial systems,* which represent the direction and force of the interaction of phenomenal systems;

- *Dimensional systems,* which model the multidimensionality of phenomenal systems;

- *Operational systems,* which define the operations of phenomenal systems;

- *Conceptual systems,* which assert the relationships of phenomenal dimensions.

As illustrated, all of the phenomenal processing systems are interdependently related; they may be entered at any stage of processing.

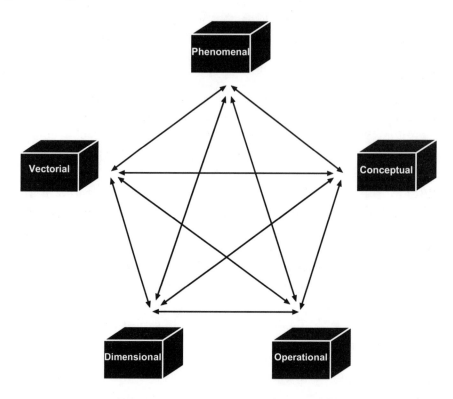

Figure 4-9. Phenomenal Processing Systems

All phenomenal processing systems are related interdependently in a similar manner; for example, the vectorial systems which emphasize organizational processing and new capital development are also related interdependently.

Generative Processing: A Case Study

The history of humankind is the history of its capital sources of economic wealth, and of the capital processing systems by which this wealth was accomplished. It is important to stress that the term "capital" is employed only to signify *what is most important or powerful in the equation for wealth.* In this context, the historic sources of wealth emphasized land, labor, and financial resources; not until the introduction of electronics technology, and relatedly, the Data Age, did information and its processing begin to make capital contributions.

In turn, the future of civilization is, emphatically, a future of new capital processing systems. Indeed, these processing systems become synonymous with capital sources of wealth. Even now we can clearly see that it is the systematic and synergistic interactions of processing systems that will account for all newly created wealth.

In the past, economists emphasized financial capital. But in the future, financial capital will be reduced to a catalytic, necessary but insufficient condition of wealth. Financial capital already accounts for *less than 15 percent* of economic productivity growth. As a "precondition" of interdependent processing, it can serve only to reduce its contribution further.

In short, our future prosperity is interdependent with our future processing. Our future processing begins with our ability to process the constantly changing requirements of our spiraling multidimensional and curvilinear environment. In shorthand terms, our future lies in our ability to process interdependently to the *nth* power!

Our entry into the realm of interdependent processing (human, organizational, marketplace, as well as informational and mechanical) and our many other forms of capital will thus account for our currently conceived wealth. And because these processing systems generate new sources of wealth, we may conclude that, together, these processing systems will account for our future growth in wealth!

In this context, the staff of Carkhuff Thinking Systems was assigned to develop an organizational processing system for software purposes. Each processor was assigned an NCD function in an interdependent processing design. As shown in Figure 4-10, the design comprised the following:

- MCD, or marketplace positioning systems;
- OCD, or organizational alignment systems;
- HCD, or human processing systems;
- ICD, or information-modeling systems;
- mCD, or mechanical tooling systems.

As usual, while all of these systems are interdependently related, they may be entered at any stage of processing.

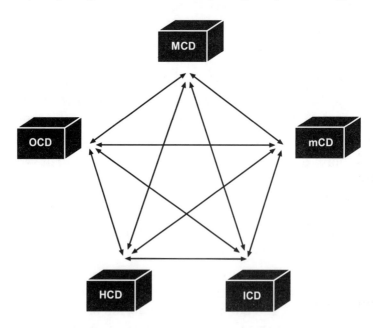

Figure 4-10. Organizational Processing Systems

Our processors were now ready to enter the organizational systems and define NCD objectives operationally. They began deductively with the highest-level MCD, or marketplace positioning systems. The objective was expressed as:

> *Marketplace positioning is discharged by corporate-capacity components enabled by organizational processing systems.*

Continuing deductively, the processors *"nested"* the OCD systems in the MCD systems. The MCD components were rotated counterclockwise to become OCD functions in a new delivery system. The OCD objective was defined operationally:

> *Marketplace-positioning functions are discharged by organizational alignment components enabled by human processing systems.*

Similarly, the HCD functions were derived deductively from the organizational alignment components. Again, the objective was defined operationally:

> *Organizational alignment functions are discharged by human processing components enabled by information modeling processes.*

Likewise, the ICD functions were derived deductively from the human processing components, with the ICD objective defined as:

> *Human processing functions are discharged by information-modeling components enabled by mechanical processing systems.*

Finally, the mCD functions were derived deductively from the information-modeling components. The mCD objective was defined operationally as well:

Information-modeling components are discharged by mechanical processing components enabled by sub-mechanical, "chaining" systems.

The processors presented the phenomenal systems deductively. Depending upon the driving functions, the delivery systems were aligned to discharge any of these phenomenal systems. Again, the phenomenal systems were entered at any stage and related to all stages.

Generative Phenomenal Modeling

Armed with the NCD-driven organizational processing systems, the processors implemented generative processing systems developmentally: interdependent processing systems processed phenomenal systems—in this instance, vectorial-organizational processing systems (see Figure 4-11). As may be noted, the organizational, or NCD, systems were *nested* in the I^5 interdependent processing systems. Keep in mind that the very process of relating to any phenomena transforms the phenomena in some ways.

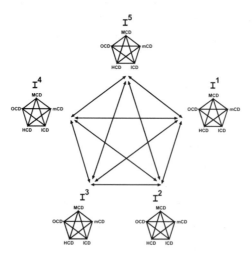

Figure 4-11. Generative Phenomenal Processing

Moving developmentally, the information relating systems transformed conceptual images of phenomena into operational images. The processes by which such transformation takes place are information relating systems, or I^1, which define phenomenal operations: functions, components, processes, conditions, standards. The I^1 processing objective was, itself, defined operationally:

> *Conceptual images are transformed into operational images by information relating systems.*

In turn, these operational images became inputs which were transformed by information representing, or I^2, into dimensional images: one-dimensional scaling, two-dimensional matrices, three-dimensional modeling, *nested*-dimensional modeling, multidimensional modeling. The I^2 processing objective was defined operationally:

> *Operational images are transformed into dimensional images by I^2 systems enabled by I^1 systems.*

Similarly, these dimensional images were transformed by individual processing systems, or I^3, into new phenomenal images: goaling, analyzing, synthesizing, operationalizing, technologizing. The I^3 processing objective was defined operationally:

> *Phenomenal images are transformed into new phenomenal images by I^3 systems enabled by I^2 systems.*

Likewise, these phenomenal images were transformed by interpersonal processing systems or I^4, into more powerful phenomenal images: goaling, getting, giving, growing, going. The I^4 processing objective was defined operationally:

> *Phenomenal images are further transformed into phenomenal images by I^4 systems enabled by I^3 systems.*

Finally, these powerful phenomenal images were transformed by interdependent processing systems, or I^5, into the most powerful phenomenal images: marketplace, organization, human, information, mechanical. The I^5 processing objective was defined operationally:

> *Powerful phenomenal images are transformed into the most powerful phenomenal images by I^5 systems enabled by I^4 systems.*

We may note that the processors presented the generative processing systems developmentally, or inductively. They could have presented them deductively. They could also have presented the systems idiosyncratically: depending upon their purpose, they could have begun at any phase of generative processing or, for that matter, any stage of phenomenal processing.

In this situation, a precondition for interdependent generative processing was expertise in the inductive and deductive model-building systems illustrated earlier. The interdependent processors could employ inductive processing systems to generate new images of phenomena, and they could employ deductive processing systems to innovate within these new images of the phenomena.

A precondition of the interdependent processing system itself was interpersonal processing. Before the individual processors could initiate the development of new images, they needed to respond accurately to the images that other processors had already expressed. In other words, before the processors could *"give"* their images, they had to *"get"* the images of others. The success or failure of interdependent processing depends upon this communicated accuracy and respect for the ideation generated by others.

Generative Processing: Applications

The major product produced by this generative processing was the NCD model itself. Sourcing from the evolving MCD model, the processors deduced the remaining models: OCD, HCD, ICD, mCD (see Figure 4-12, on the following page). Each lower-order processing system was *nested* in a higher-order processing system. All were *nested* in the MCD model. Moreover, each lower-order model was deduced, or derived, by rotating components and processes counterclockwise from its higher-order source. All were transformed in identity so that the level of the model's functions could be derived. This was deductive intelligence in operation.

In turn, each lower-order processing was rotated clockwise to generate higher-order systems. In other words, the NCD model operated inductively: mCD, ICD, HCD, OCD, MCD. This was inductive intelligence in operation. Both inductive and deductive intelligence are incorporated in interdependent generative processing, as we will see.

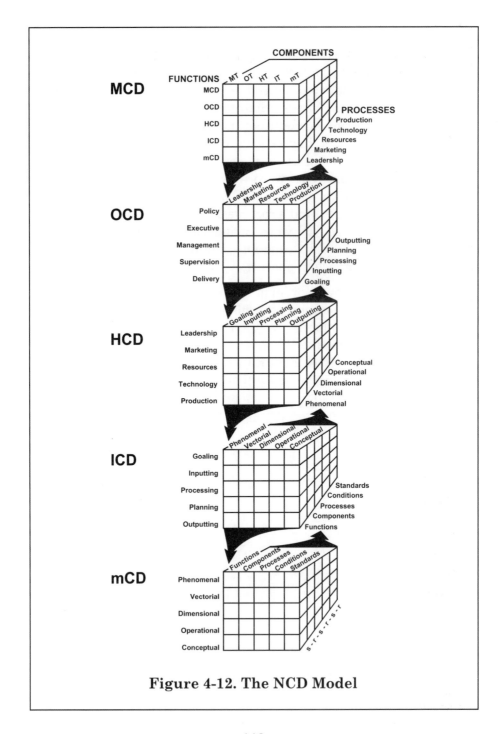

Figure 4-12. The NCD Model

Finally, the processors culminated their work in the generative processing systems. Crossing interdependent processing systems were arranged developmentally, or inductively: I^1, I^2, I^3, I^4, I^5. In turn, the organizational phenomenal functions were derived deductively: MCD, OCD, HCD, ICD, mCD (see Table 4-1).

Table 4-1. The Human Intelligence Matrix

Organizational Phenomena	Generative Processing Systems				
	I^1	I^2	I^3	I^4	I^5
MCD					
OCD					
HCD					
ICD					
mCD					

The processors viewed the interdependent processing systems as *"possibilities leadership"* systems.* In turn, they viewed the organizational systems as *"possibilities organization"* systems.** They understood that the interdependent processing skills enable the *"possibilities leader"* to accomplish the new capital development functions of the *"possibilities organization."*

---

* Carkhuff, R. R., and Berenson, B. G. *The Possibilities Leader.* Amherst, MA: HRD Press, 2000.
** Carkhuff, R. R., and Berenson, B. G. *The Possibilities Organization.* Amherst, MA: HRD Press, 2000.

The processors also viewed generative processing systems in relation to all phenomena (see Table 4-2). Here the I^5 systems were dedicated to discharging functions in all phenomena: physical-universal, natural-environmental, biochemical, community/human, and organizational/human. We have already illustrated I^5 in organizational and community areas. What does it mean to I^5 in the other natural phenomenal areas? Does the experience come alive for us? Can we represent their operations dimensionally? Can we process with them—individually, interpersonally, interdependently?

Table 4-2. Matrix for Human Intelligence With All Phenomena

Phenomena	Generative Processing Systems				
	I^1	I^2	I^3	I^4	I^5
Physical-Universal					
Environmental					
Biochemical					
Community/ Human					
Organizational/ Human					

Corporate Productivity Upgrade[*]

The recent focus upon inter-enterprise operations is creating the requirements for common, deductive, and interdependent information environments. Within these information environments, entire constellations of business organizations interrelate to redefine themselves continuously and dynamically.

Until the spiraling requirements of information change forced inter-enterprise management upon us, independent organizational decision-making was the model. Today, if we ignore inter-enterprise management, we exclude partners, suppliers, vendors—and especially their business offerings—from the most critical and formative moments of strategic processing.

Relatedly, the traditional consulting functions will be displaced by new products and services that enable a self-organizing business environment. This business environment will be based upon the principles of interdependence and human processing within the most powerful modeling and management environments.

In this context, let us consider the situation of a specific company. Marketing and common business objectives drive three of its operations: CAD-CAM, data management, and a services organization. Currently, the company's focus is upon product life-cycle management. Product life-cycle management aligns directly to decision making at the management level.

From another perspective (see Table 4-3), we may scale the levels of customer business operations confronting the company:

---

[*] Carkhuff, C. J., and Carkhuff, R. R. *Corporate Productivity Upgrade.* McLean, VA: Carkhuff Thinking Systems, 1999.

- Standard applications training,
- Boutique applications,
- Product life-cycle management,
- Systems integration,
- Strategic processing.

As may be noted, the company's concentration is upon the first three levels, culminating in product life-cycle management. The company does not address the higher levels of customer business operations: systems integration and strategic processing.

Table 4-3. Levels of Customer Business Operations Addressed by Corporation

LEVELS OF
CUSTOMER
OPERATIONS

5. **Strategic Processing**

4. **Systems Integration**

3. **Life-cycle Management**

2. **Boutique Applications**

1. **Standard Applications Training**

The limitation of products and services to product life-cycle management has powerful business implications. Large customers represent the core of many business constellations. Policy- and executive-level alignment around strategic and integrated decision-making will account for most of the variance in the company's success. The company needs products and services in these spaces.

The Story of CPU

For Carkhuff Thinking Systems, the story begins in the manufacturing marketplace with EPD: electronic product definition. EPD initiated a powerful business movement, one that has yet to realize its business potential. CTS's study of sources of new capital development led to the identification of new sources of variance in EPD-driven efforts:

- Common models of products that enable concurrent efforts in design and development;

- Resulting maps of processes that move companies beyond life-cycle management—maps that enable resource positioning and direction;

- Collaborative human-processing environments that actualize the human contribution.

The message of EPD has been obfuscated by competition and a commoditizing business mindset, instead of used to identify major sources of variance, and to transfer major functions from the engineering and manufacturing worlds to inter-enterprise organizations and markets.

Corporate productivity upgrade, or CPU, is a business solution designed to replace expensive investments in strategic consulting and systems integration. The CPU initiative enables customers to develop interdependent and inter-enterprise solutions while simultaneously facilitating corporate cultural change or cultural capital development.

CPU is designed to effect interdependent project and process lifestyle management. CPU is composed of three major sets of processing operations:

➡ **INCOME**   *Interdependent common-object-modeling environment,* a software solution for describing projects and initiatives interdependently.

➡ **COMPASS**   *Concurrent organization-mapping and process-alignment software solutions,* a deductive corporate map assembled from common objectives.

➡ I⁵   *Interdependent processing systems* which empower human capital and help transform it into business capital.

The CPU model, shown in Figure 4-13, presents the operational dimensions of CPU:

- *Functions*—Describing, predicting, and prescribing;
- *Components*—Software, methods, and training;
- *Processes*—Modeling, mapping, and processing.

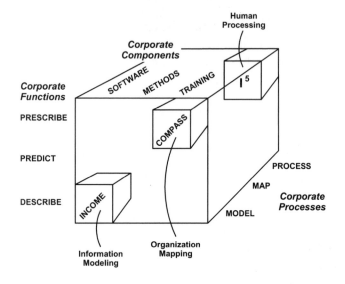

Figure 4-13. The CPU Model

As we can see, the critical processing operations are nested as *"source cells"* within the CPU model:

➡ **INCOME**   *Interdependent common-object-modeling environments,* which model common objectives into information objects to discharge accurate describing functions for inter-enterprise components.

➡ **COMPASS**   *Concurrent organization-mapping and process-alignment software solutions,* which enable all resources to identify their current positioning, relationships, and directions interdependently; they also improve prediction functions for concurrent initiatives in relation to market experiences and value and supply chains.

➡ I^5 *Interdependent processing systems,* which empower human capital by information relating, information representing, and generative processing—individual, interpersonal, and interdependent—in a collaborative information environment.

We may summarize the processing operations as illustrated:

➡ **INCOME**   Information modeling,

➡ **COMPASS**   Organization mapping,

➡ I^5 Human processing.

The CPU model was derived from the new science of possibilities (see Figure 4-14). As may be noted, the CPU model itself *nests* within the new capital development models of the vectorial information-modeling components. By our rotating the components to discharge the functions of describing, predicting, and prescribing (relating), the O-H-I delivery systems are transformed through H-I-O delivery systems to I-O-H delivery systems: information software, organizational methods, and human systems training. Relatedly, we align the processes to enable I-O-H information components to discharge describing, predicting, and prescribing (relating) functions: I^1, I^2, I^3.

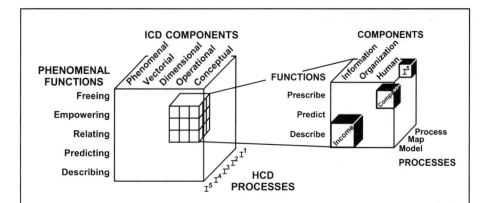

**Figure 4-14. CPU Model Derived From
The New Science of Possibilities**

We began by building upon the cornerstone of electronic product definition, or EPD, which drove the CAD-CAM applications in manufacturing. With iterations of applications, we had now defined the prepotent sources of productivity: electronic processing definition, or EP⁵D. EP⁵D interdependently relates all processing systems within which EPD operates: MCD, OCD, HCD, ICD, mCD.

In summary, CPU is designed to free company assets to concurrently transform resources into capital assets while upgrading the corporate culture to provide true comparative advantage. Currently, company assets are frozen by *"budgets, blocks,* and *locks"*—planning locks. Utilizing current resources, companies can make significant productivity gains by empowering and freeing current assets with CPU.

IN TRANSITION

Generative processing is the model for human intelligence. It enables *"possibilities leaders"* to develop the interdependent processing systems required to generate more powerful images of DNA, NCD, or nature itself. It enables the human processors to be not only intelligent but intentional—to accomplish what they set out to accomplish. And because interdependent processing is continuous, with images constantly changing, processors are able to fulfill their intentions in a form even better than originally envisioned.

The relationship between intelligence and intentionality is synergistic: each grows as the other grows. Intelligence yields intentional hypothesis-testing. Intentionality builds intelligence from the results of these tests.

In a different context, then, we might dedicate phenomenal components to discharging I^5 functions (see Table 4-4). In this case, we are committed to developing generative processing systems or skills by processing organizational processing systems: MCD, OCD, HCD, ICD, mCD. In other words, the NCD systems accomplish the development of generative intelligence: I^5, I^4, I^3, I^2, I^1. Again, we may implement phenomenal intelligence inductively, deductively, or idiosyncratically.

Table 4-4. Matrix for Phenomenal Intelligence

Generative Processing Functions	Phenomenal Processing Systems				
	MCD	OCD	HCD	ICD	mCD
I^1					
I^2					
I^3					
I^4					
I^5					

In viewing the generative processing systems shown in Figure 4-15, we may see any of the phenomenal systems (1 through n) as *nested* within the I^5 systems. Again, all processing systems, phenomenal as well as human, are interdependently related—which means that we can enter any system at any stage and relate to all systems at all stages.

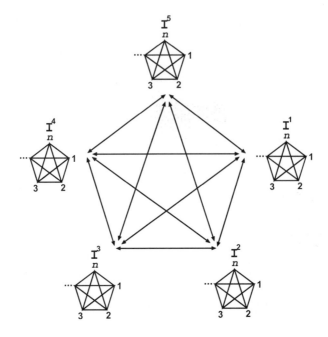

Figure 4-15. Generative Processing Systems

Keep in mind that just as the phenomenal systems may be *nested* in the generative systems, so may the generative systems be *nested* in the phenomenal systems. We may conclude that each interdependent processing system may be *nested* in the other, depending upon our purposes or functions.

In the final analysis, we may relate the models for generative and phenomenal intelligence. Basically, any combination of components and functions of any model may relate to any combination of functions and components of any other model; thus, we may tailor our designs for generative intelligence.

In transition, generative intelligence empowers us to engage in continuous processing and rapid prototyping by relating our inductive and deductive processes interdependently and synergistically, as illustrated in the NCD design below (Figure 4-16). We may compare this illustration of generative intelligence with our previous illustrations of inductive and deductive intelligence (see Figures 2-26 and 3-26). Each level is more powerful than the previous level in generating degrees of freedom and phenomenal information.

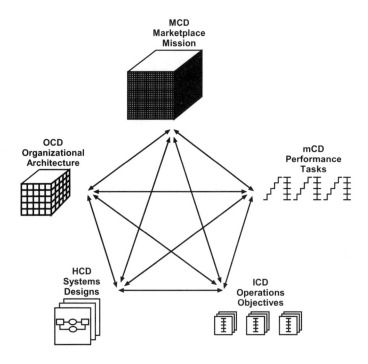

Figure 4-16. Generative-Intelligence
Illustration of NCD

Generative intelligence empowers us to source the growth and, indeed, the life of any phenomena. It empowers us to process, if not govern, our own destinies with intelligence and intentionality. It is an accelerator of evolution. All phenomena, including especially humans, no longer have to endure the painfully slow, accidentally random, and often destructively alien coding of evolution.

5 Changeability and Hybrid Modeling

Dimensional intelligence is the prerequisite for generative intelligence. With dimensional intelligence—whether deductively derived or developmentally induced—generative processing is possible. Without dimensional intelligence, generativity is not possible. With both deductively derived and developmentally induced dimensional intelligence, the highest levels of generative processing are possible.

Dimensional intelligence simply means that we have defined the phenomena—whatever phenomena concern us—multidimensionally: their functions, components, processes; their conditions and standards. Dimensional intelligence allows us to relate all of these dimensions interdependently. Moreover, dimensional intelligence empowers us to generate new and more powerful dimensions, and thus, new and more powerful phenomena. The generation of these new and more powerful dimensions is the core of generative processing.

All generative processing may emanate from this convergence of factors and vectors as a source (see Figure 5-1). As illustrated below, the generative initiatives may go out in any direction. Indeed, they may relate to all operations of possibilities science: phenomenal freeing functions, information-modeling components, interdependent processing systems, freeing and changeable conditions, and standards.

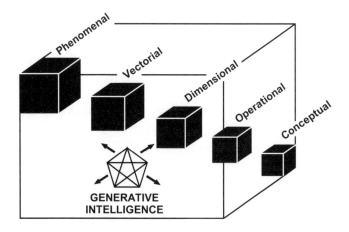

Figure 5-1. Interdependent Generative Intelligence

THE NEW MONOLITHIC IDEA

In introducing the Data Age, Jack Kilby developed the *"The Monolithic Idea"* in honor of the monolithic integrated electronic circuit. The Monolithic Idea stood for astonishing reductions in size and elevated integrations in quality of all parts of an electronic circuit. Today the invention is better known as the semi-conductor chip, and is commonly referred to as the microchip. The human implications of this technological innovation were, and still are, profound.

Not only did the "chip" pervade all forms of electronic products, but by making computers universally available, it brought about the end of Industrial Age dominance. In so doing, it ushered in an age in which micro-electronic and digital technologies converged upon a common mission for civilization: the unimpeded flow of global information.

With the development of the Data Age and the extension of information technologies, the requirements for human processing have been elevated. And with this elevation has come the generation of a breakthrough idea: the integration of all processing technologies—*the new Monolithic Idea* for processing systems. In a situation where information systems have already witnessed diminishing returns in comparative advantage or competitive edge, the new Monolithic Idea, or the new M.I., promises to empower these systems to actualize their contributions, just as the information systems empowered the mechanical systems of the industrial age. In the new M.I., all known processing systems interact with one another: marketplace, organizational, human, information, and mechanical.

The new M.I. differs from the old M.I. on one critical dimension: changeability versus stasis. The old M.I. presented an integrated circuit that was static in form and function. The new M.I. presents an integrated circuit that is dynamic in form and function. Quite the opposite of the disruption that probabilities science projects, changeability is what holds phenomena together: changeability is the connective bonding tissue for all phenomena. Indeed, the new M.I. is changeability!

We may represent the integrated processing systems for possibilities science as shown in Figure 5-2. As we can see, every processing system interacts with every other processing system. In point of fact, this new M.I. is a factored representation of multidimensional space. The multidimensional vectors involved are, themselves, factored from convergent points in space and time in multidimensional, curvilinear space. In other words, the new M.I. is a human representation of currently natural and continuously changing phenomena. However, because it represents an integration of processing systems, the new M.I. presents the potential of humans creating their own future environments. Thus, future representations of integrated processing may take very different forms.

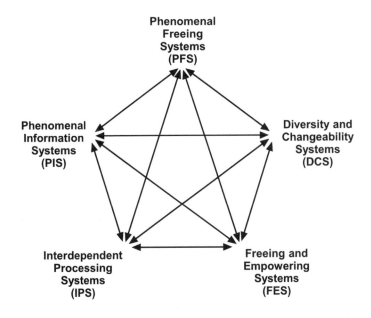

Figure 5-2. Possibilities Science Processing Systems

As may be noted, every entity in possibilities science is a processing system. Moreover, every processing system is related interdependently to every other processing system. Again, it does not matter where we enter such a system. By processing interdependently, we can establish the nature of the other entities in the processing system. Indeed, for our purposes, we may recognize the entire system as a *"possibilities object."*

From another perspective, we may telescope the processing systems to focus upon the mission of the dominant system. That is to say, each processing system operates within a dominant or prepotent processing system (see Figure 5-3). Note that as the focus of processing evolves, the positioning of the processing systems shifts.

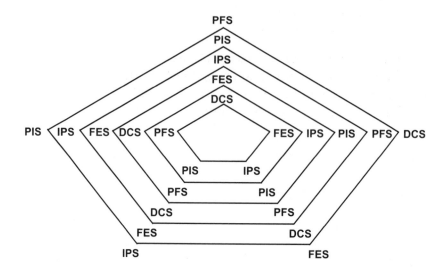

LEGEND:

PFS = Phenomenal Freeing Systems
PIS   = Phenomenal Information Systems
IPS   = Interdependent Processing Systems
FES = Freeing and Empowering Systems
DCS = Diversity and Changeability Systems

Figure 5-3. A Telescoped Image of Possibilities Processing System

In this context, the processing systems are ordered in an hierarchical manner: the lower-most systems operate within the upper-most system. We may view them as delivery systems. The overall purpose of possibilities science is to discharge the phenomenal freeing functions. All other processing systems align with these functions:

- Phenomenal freeing functions (PFS),
- Phenomenal information components (PIS),
- Interdependent processing systems (IPS),
- Freeing or empowering conditions (FES),
- Diverse and changeable standards (DCS).

As illustrated, other processing systems may be placed in ascendancy and the remaining systems aligned. For example, we may rotate deductively, or counterclockwise, to accomplish the development of phenomenal information. In this instance, all other processing systems align with these new functions:

- Phenomenal information functions (PIS),
- Interdependent processing components (IPS),
- Freeing and empowering processes (FES),
- Diverse and changeable conditions (DCS),
- Phenomenal freeing standards (PFS).

The phenomenal freeing functions are shifted to standards in this realized delivery system. Basically, we measure whether or not the re-aligned system has accomplished the original purposes.

In a similar manner, we may rotate inductively, or clockwise, to serve other delivery functions. For example, we may rotate inductively to accomplish the development of diverse and change-able standards. In this instance, all other processing systems align with these new processing functions:

- Diverse and changeable functions (DCS),
- Freeing and empowering components (FES),
- Interdependent processing systems (IPS),
- Phenomenal information conditions (PIS),
- Phenomenal freeing standards (PFS).

The fundamental assumption of the new M.I. is this:

> *Processing systems account for all new responses in rapidly changing marketplace environments by systematically integrating them.*

Due to accelerated use of processing systems, 95 percent of the *breakthrough* ideas in the history of humankind have occurred in the last decade. It is a small step to recognizing that all of our future *breakthroughs* reside in our processing potential. This potential is a function of developing and integrating all current and future processing systems.

The fundamental assumption of the new M.I. generates a number of processing principles that account for the sources of variance or effect between systems. In all, there are five principles:

- *Integrated processing systems generate synergistically to process one another's growth.*

- *Integrated processing systems generate to actualize the contributions of all known sources of variance.*

- *Integrated processing systems generate the discovery of new sources of changeability and, thus, new responses.*

- *Integrated processing systems generate new environmental requirements based upon the creation of new responses.*

- *Integrated processing systems generate the discovery of new processing systems as totally new sources of changeability.*

In short, the integration of all known processing systems generates all new responses, including the sources of these responses.

The processing principles also generate a number of powerful corollaries that account for the sources of effect within systems. Those corollaries are as follows:

- *Each level of processing systems constitutes a source of effect within which other levels of processing systems operate.*

- *Each level of processing systems frees the contributions of every other level of processing systems.*

- *Each level of processing systems is required to process new content.*

- *Each level of processing systems is empowered to create new content.*

- *Each level of processing systems is empowered to create new processes.*

In short, each level of processing systematically generates every other level's processing and, thus, its own sources of change.

In summary, the new M.I. integrates all known processing systems dedicated to the generation of new responses. Every processing system and each level of processing impose requirements for processing upon every other processing system and level. In doing so, the integrated processing systems systematically generate their own sources of change and, thus, requirements. In short, the new M.I. generates the integration of processing systems by which humans, organizations, and communities can create their own destinies.

Interdependent Generative Processing: A Case Study

Employing generative intelligence, the processors at Carkhuff Thinking Systems processed interdependently to develop a new scientific paradigm (see Figure 5-4). In this paradigm, all dimensions of science are processing systems and all processing systems are related interdependently.

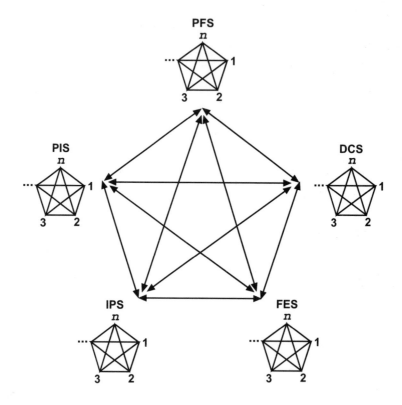

Figure 5-4. Possibilities Processing Systems

As shown in Figure 5-4, all processing systems have other processing systems *nested* within them. These may be both interdependent and substantive processing systems. For example, the phenomenal freeing systems (PFS) could *nest* both I^5 and substantive processing systems; such *nesting* would mean that the possibilities-science processor had expertise in both types of processing systems.

The processors defined the objective at this PFS station operationally:

> *Phenomenal freeing functions are discharged by interdependent processing components enabled by substantive processing systems.*

They knew that I^5 and substantive processing systems could *nest* in all of the other science stations as well:

- PIS, or phenomenal information systems;
- IPS, or interdependent processing systems;
- FES, or freeing and empowering systems;
- DCS, or diverse and changeable systems.

This meant that the processors' abilities to engage generatively in possibilities processing was contingent upon their skills in interdependent processing systems: I^1, I^2, I^3, I^4, I^5. In turn, their applications of interdependent processing systems were contingent upon their expertise in the substantive processing specialties involved.

The substantive processing systems may incorporate any system with which we are familiar. In this case, they included the NCD systems: MCD, OCD, HCD, ICD, mCD. They also included some of the higher-order systems from which NCD systems were derived:

- CCD—Cultural capital development,
- ECD—Economic capital development,
- GCD—Governance capital development.

Finally, the substantive processing systems included higher-order systems:

- Physical-universal systems,
- Ecological-environmental systems,
- Biochemical systems.

All systems were phenomenal processing systems.

It is important to emphasize that all entities were rotated deductively and inductively, depending on their delivery systems. Thus, the possibilities processing systems were rotated as follows: PFS, PIS, IPS, FES, DCS. Similarly, and simultaneously, the interdependent processing systems—I^1, I^2, I^3, I^4, I^5—were rotated. Finally, the substantive processing systems—for example, MCD, OCD, HCD, ICD, mCD—were rotated simultaneously as well. Again, the possibilities generated by the processors were contingent upon their skills in possibilities processing, interdependent processing, and substantive processing.

Foremost among the generative products of this possibilities processing was the new scientific model itself: the new science of possibilities.

The possibilities model is presented deductively in Figure 5-5; in this case, it is labeled "Possibilities Phenomena" and helps to illustrate the standards of possibilities science. As we can see, the components are driven from the phenomenal level down: phenomenal, vectorial, dimensional, operational, conceptual. At the same time, the interdependent processes may be driven from the interdependent level down: interdependent, interpersonal, individual, informational. Remember, deductive modeling is powerfully leveraged when phenomena are already known to us.

At this point, let us review the possibilities conditions and standards. Again, the possibilities conditions are the environments within which the phenomena are *nested*. As may be noted, the conditions have their own unique operations:

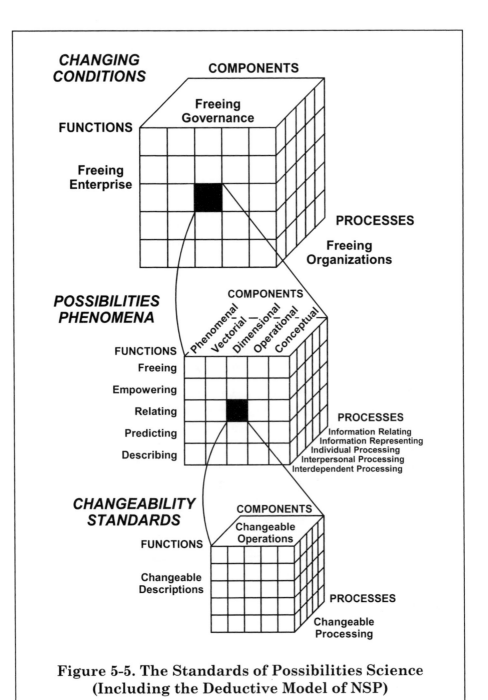

**Figure 5-5. The Standards of Possibilities Science
(Including the Deductive Model of NSP)**

functions, components, and processes. If the conditions are freeing, then the possibilities science can work its magic. Free-enterprise economics maximizes free choice for individual corporate entities. Free and direct democratic governance maximizes free choice for individual human phenomena. Free organizations maximize free choice in aligning resources for continuously evolving marketplace and governance objectives. Within the context of these conditions, the freeing functions of possibilities science may be served.

We can readily see possibilities conditions in the environment of our science of possibilities and the phenomena it generates: the freeing conditions of free-enterprise capitalism; the freeing conditions of direct-representation democracies; the freeing conditions of self-organizing, *"possibilities"* organizations. These freeing conditions not only allow possibilities to be generated, but also enable them to be generated; moreover, in interdependent processing with the phenomena, the conditions themselves contribute to generating possibilities.

Possibilities standards are the measures of performance required to meet the freeing functions. Again, the standards have their own unique operations. Whereas the probabilities measures stress uniformity, the possibilities measures emphasize diversity. These measures maximize the range of performance. Ultimately, diversity is transformed into measures of changeability as the phenomena evolve in continuously changing forms. The standards are simply measuring what the new science is all about: possibilities.

Interdependent Generative Processing: Hybrid Modeling

At the human-systems level, we say, *"The answer is empowerment—what is the question?"* At the organizational-systems level, we say, *"The answer is hybrid modeling—what is the question?"*

Hybrid modeling is simply the presence of different operations dedicated to the same or allied functions. It emphasizes the transfer of operations by holding functions constant while varying the other operations: components, processes, contexts, conditions. Hybrid modeling is most often a transitional design incorporating current and future operations simultaneously. Such modeling is most useful in these times of spiraling change and elevated requirements. It ensures the success of our operations' movement toward greater productivity.

By way of illustration, we will explain in this section how we at Carkhuff Thinking Systems employed hybrid modeling in our interdependent generative processing.

First, we turned to our OCD model and mapped into its functions (policy, executive, management, supervision, delivery) and components (leadership, marketing, resource integration, technology, production). Next, we inserted the *nested* processing systems into the OCD Master Matrix[*] (see Table 5-1). These processing systems were inserted on the diagonal to represent *"the z dimension,"* or third dimension, of processing. They are as follows:

- S-MP-R marketplace positioning systems,
- S-OP-R organizational alignment systems,
- S-P-R human processing systems,
- S-O-R information-modeling systems,
- S-R mechanical tooling systems.

---

[*] Carkhuff, C. J., and Carkhuff, R. R. *The OCD Matrix.* McLean, VA: Carkhuff Thinking Systems, Inc., 1995.

139

As may be noted, the hierarchy of processing systems are rotated through the *source cells* of OCD.

Table 5-1. The OCD Master Matrix

COMPONENTS

FUNCTIONS	Leadership	Marketing	Resources	Technology	Production
Policy	S-MP-R				
Executive		S-OP-R			
Management			S-P-R		
Supervision				S-O-R	
Delivery					S-R

This rotation of *nested* processing systems gave us the two-dimensional image of multidimensional processing shown in Figure 5-6. As illustrated, the 2D images were transformed into *"z dimension"* source-cell images: S-MP-R, S-OP-R, S-P-R, S-O-R, S-R. We then defined the cell operations as follows:

Policy-positioning functions are discharged by leadership components enabled by goaling processes.

Executive-aligning functions are discharged by marketing components enabled by inputting processes.

Management-systems-design functions are discharged by resource-integration components enabled by generative processing systems.

Supervisory modeling functions are discharged by technology components enabled by planning processes.

Delivery-performance functions are discharged by production components enabled by outputting processes.

To summarize:

> *All levels of organizational functions (PEMSD) are discharged by rotated areas of components (LMRTP) enabled by rotated phases of processing (GIPPO).*

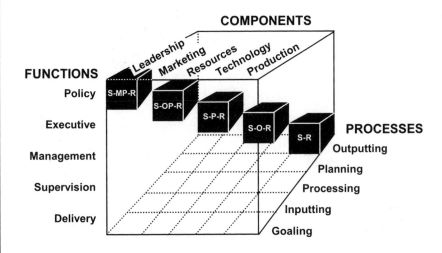

Figure 5-6. *Nested* **Processing Systems in OCD Model**

We were now able to model multidimensionally using our two-dimensional OCD matrix. For example, we could align our marketing component at all functional levels of the organization, as shown in Table 5-2. Notice that each level of the marketing component emphasizes one of the *nested* processing systems: S-MP-R, S-OP-R, S-P-R, S-O-R, S-R.

Table 5-2. Aligning Marketing Components in OCD Matrix

COMPONENTS

FUNCTIONS	Leadership	Marketing	Resources	Technology	Production
Policy		S-MP-R			
Executive		S-OP-R			
Management		S-P-R			
Supervision		S-O-R			
Delivery		S-R			

Again, this external rotation of *nested* processing systems produces a two-dimensional image of multidimensional processing (see Figure 5-7). The objectives of the marketing-components operations may be defined as follows:

Policy-positioning functions are discharged by marketing components enabled by goaling processes.

Executive-aligning functions are discharged by marketing components enabled by inputting processes.

Management-systems-design functions are discharged by marketing components enabled by generative processing systems.

Supervisory modeling functions are discharged by marketing components enabled by planning systems.

Delivery-performance functions are discharged by marketing components enabled by outputting systems.

To summarize:

All levels of organizational functions (PEMSD) are discharged by the marketing components enabled by all phases of organizational processing (GIPPO).

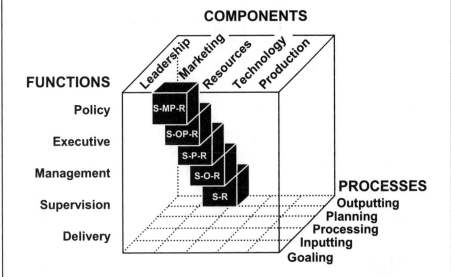

Figure 5-7. Marketing Components Aligned in OCD Model

We also could illustrate aligning management functions through all component areas of the organization, as shown in Table 5-3. Note that each area of the management level emphasizes one of the *nested* processing systems: S-MP-R, S-OP-R, S-P-R, S-O-R, S-R.

Table 5-3. Aligning Management Functions in OCD Matrix

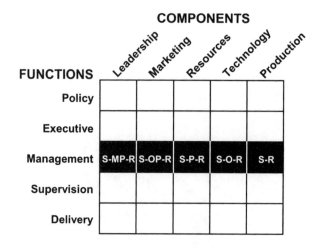

COMPONENTS

FUNCTIONS	Leadership	Marketing	Resources	Technology	Production
Policy					
Executive					
Management	S-MP-R	S-OP-R	S-P-R	S-O-R	S-R
Supervision					
Delivery					

Once again, this external rotation of *nested* processing systems produces a two-dimensional image of multidimensional processing (see Figure 5-8). The objectives of the management operations may be defined as follows:

Management-systems-design functions are discharged by leadership components enabled by goaling processes.

Management-systems-design functions are discharged by marketing components enabled by inputting processes.

Management-systems-design functions are discharged by resource-integration components enabled by processing systems.

Management-systems-design functions are discharged by technology components enabled by planning processes.

Management-systems-design functions are discharged by production components enabled by outputting processes.

To summarize:

The management-systems-design functions are discharged by all organizational components (LMRTP) enabled by all phases of organizational processing.

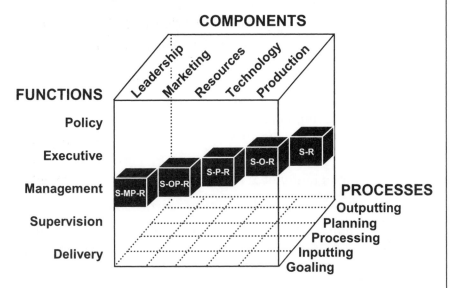

Figure 5-8. Management Functions Aligned in OCD Model

It is important to stress that these simple demonstrations yield perspective on the Master Matrix. All dimensions may be rotated to represent complex organizational phenomena.

Interdependent Generative Processing: Applications

The manufacturing company, a major producer of large and complex products, was positioned in the marketplace as an innovative standard setter. But its people soon realized that its operations had reached a level of unmanageable complexity. Here is what they did. They formed teams of exemplary engineers. Utilizing the principles of concurrent engineering, they *"factored out"* all of the major operations of product building. They formed units representing these major functions; then they factored within these major operations the primary functions upon which all unit components and processes had to bear.

In effect, they formed a *"team of teams."* They called it "concurrent process engineering." Here is the approach the team used. Their mission was to develop Total Product Modeling. Their goals were as follows:

1. To implement global concurrent engineering;

2. To implement collaborative organizational processes;

3. To change their engineering processes from serial to parallel and, ultimately, to rapid prototyping—concurrent processing.

Utilizing cross-functional analysis, the concurrent-engineering team addressed the ingredients of their process-centric image (Figure 5-9). Five major factors bore upon the process: co-orchestrated organizational change to keep up with changes in the marketplace; cooperative organizational teaming methods (such as their demonstration of process reengineering); collaborative human-processing methods; communicative information technologies; and coordinated mechanical tooling, including information connectivity.

146

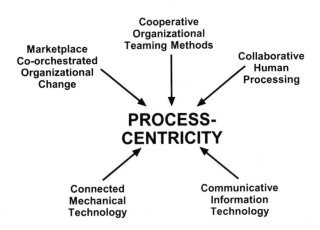

Figure 5-9. The Processing Ingredients

The concurrent-engineering team believed they had an urgent need to technologize all of these areas in order to accomplish their mission of Total Product Modeling. We may view the focus of the team's technologizing effort in Table 5-4. As shown, their intent is to move upward from their current connectivity and communication to higher levels of objectives: collaboration, cooperation, co-orchestration. As also shown, each level of effort is defined by a level of new capital development: marketplace, organization, human, information, mechanical.

Table 5-4. Technologizing Objectives

5 Co-orchestration (Marketplace)

4 Cooperation (Organization)

3 Collaboration (Human)

2 Communication (Information)

1 Connectivity (Mechanical)

The vision of the basic model is represented in Figure 5-10. As may be noted, the functions are marketplace functions: organizational change to orchestrate alignment with continuous repositioning in the marketplace. In turn, the components are organizational teaming processes dedicated to cooperative work efforts such as the process reengineering. Finally, the processes are human processing systems dedicated to interdependent and collaborative processing. We may formulate this mission as the organizational capital development (OCD) mission:

> *Co-orchestrated organizational change functions are accomplished by organizational teaming components enabled by interdependent human processing.*

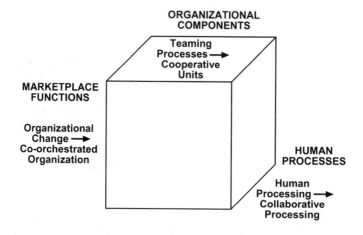

ORGANIZATIONAL
COMPONENTS

Teaming
Processes ➝
Cooperative
Units

MARKETPLACE
FUNCTIONS

Organizational
Change ➝
Co-orchestrated
Organization

HUMAN
PROCESSES

Human
Processing ➝
Collaborative
Processing

Figure 5-10. The Organizational Capital Development Mission

When we rotate our organizational teaming components deductively, or counterclockwise, they become the functions of our human capital development (HCD) mission (see Figure 5-11). In turn, the human processing systems become the components dedicated to accomplishing the teaming functions.

Finally, new information processes are introduced. They accomplish the communication that enables the cooperative human-processing components to accomplish the collaborative organizational teaming functions. We may formulate this human capital mission as follows:

Cooperative organizational teaming functions are accomplished by collaborative human-processing components enabled by information-communication processes.

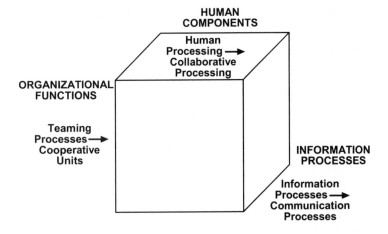

Figure 5-11. The Human Capital Development Mission

When we rotate our interdependent human-processing components deductively, they become the functions of our information capital mission (see Figure 5-12). In turn, information-communication processes become the components dedicated to accomplishing our human processing functions. Finally, new mechanical processes are introduced. They accomplish the connectivity that enables the information communication to accomplish the human processing functions. We may formulate this information capital mission as follows:

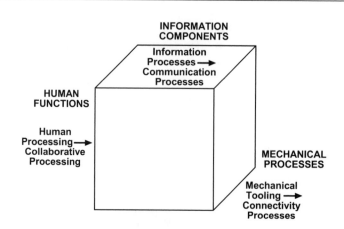

Figure 5-12. The Information Capital Development Mission

Collaborative human-processing functions are accomplished by information-communication components enabled by mechanical connectivity.

This is the basic paradigm for the possibilities organization and organizational change. Change is a function of new capital development systems:

- MCD—Marketplace capital development,
- OCD—Organizational capital development,
- HCD—Human capital development,
- ICD—Information capital development,
- MCD—Mechanical capital development.

As we can see in Figure 5-13, each new capital development system is nested in the marketplace system: MCD, OCD, HCD, ICD, mCD. New capital development (NCD) is the source of all organizational change:

NCD ⟶ Change

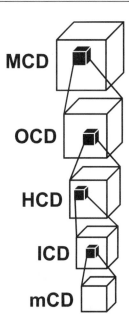

Figure 5-13. The Possibilities Organization as a Function of the New Capital Development System

The concurrent-engineering team was eminently successful in bringing about desirable change. The team subtracted more than a year from product development and established a 30 percent reduction in costs.

In this context, the Hybrid Model enabled the modeling of multiple paradigms. Such modeling technology, called *"paradigmetrics,"* is based upon the five core processing systems: S-MP-R, S-OP-R, S-P-R, S-O-R, S-R. A major requirement for the application of paradigmetrics is the organization of the core processing systems into paradigms representing the complexity of the focal organizational experience (see Figure 5-14). As may be noted, the processing systems are rotated to generate new processing *"worlds"* for each function: marketplace, organization, human, information, mechanical.

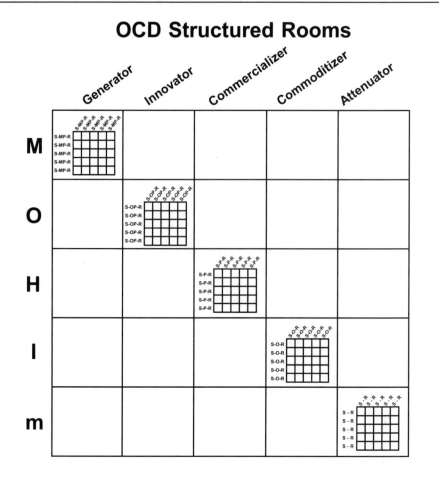

Figure 5-14. The Hybrid Model

This means that each *"world"* is identified by its function:

- Marketplace: MOHIm.
- Organization: MOHIm.
- Human: MOHIm.
- Information: MOHIm.
- Mechanical: MOHIm.

We may view illustrations of these personalized paradigms in Figure 5-15, and express them in the following way:

- Marketplace systems (MOHIm) dedicated to market-place functions (PEMSD);

- Organization systems (MOHIm) dedicated to organization functions (LMRTP);

- Human systems (MOHIm) dedicated to human functions (GIPPO);

- Information systems (MOHIm) dedicated to information functions (PVDOC);

- Mechanical systems (MOHIm) dedicated to mechanical functions (FCPCS).

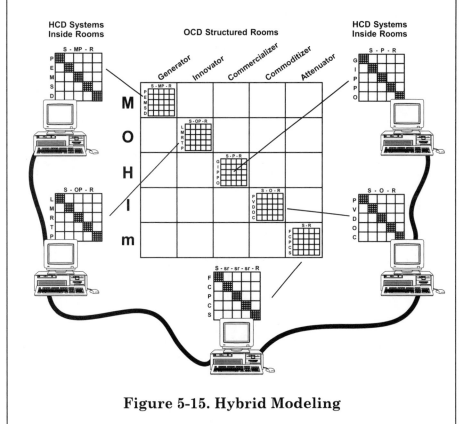

Figure 5-15. Hybrid Modeling

Thus, each of these resulting paradigms generates its own unique contributions to the entire enterprise. Because of the rules set forth by the interrelations of the core processing systems, each paradigm is capable of a near-infinite variety of responses to the internal and external environment. These paradigms are organized into a matrix to represent a deductive hierarchy of paradigms. Each paradigm is then defined as a hierarchy of the full processing potential.

The Hybrid Model's processing potential is mechanically calculated and recalculated to align with the requirements of a changing environment or marketplace. Users select a paradigm or a hybrid paradigm, and then personalize it to their unique functions. The multiparadigmetric design automatically relates any one paradigm to all other paradigms, and vice versa. The relationships of paradigms are always responding to a changing environment; the changing environment is being processed by the multiple paradigms to create simultaneous, multiple, validated worlds.

The implications for marketplace positioning are profound. As the global economy becomes integrated and interdependent, marketplace positioning likewise becomes interdependent and integrated. Financial capital continues to obey the basic laws of economics: financial resources are continuously invested and reinvested in the most valued uses. An interdependent global economy funds capital migrating freely to realize the best returns, thus making the economy more resilient. In short, market capital will "chase" the most powerfully positioned companies into the future, rather than allow them to fall into the past.

Managing marketplace positioning projects the business and its enterprise networks in the marketplace. Positioning is futuristic in orientation. Managing marketplace positioning emphasizes preparing as-yet-unborn generations of business opportunities for continuous repositioning in the marketplace. We label this process "marketplace capital development," or

"MCD," because it dictates the future success of our companies.

We can best understand the actual nature of positioning by considering its sources. Let us consider the sources of positioning for two manufacturers that have the same positioning in the marketplace: Company A and Company B. Both are positioned as manufacturing companies dedicated to delivering high-quality products at the lowest costs. Both are leaders in the industry, each with remarkable track records. Both are dedicated to market share for the business and profit share for their customers.

What makes these two companies different is their dedication to different paradigms. One company employs the probabilities paradigm that dominated the Industrial Age: *the continuous planning of operations to achieve fixed goals.* The other company uses the possibilities paradigm now emerging in the Information Age: *the continuous processing of operations dedicated to continuous evolving goals.*

Company A is planning-centric within the probabilities paradigm, as shown in Figure 5-16. As we can see, once the goals are fixed, the planning systems emanate out to the various operations. Moreover, because the goals are fixed, the planning systems are driven by politically sensitive, internal negotiations leading to consensus among technical experts.

**Figure 5-16. The Planning-Centric
Probabilities Paradigm**

This planning-centric paradigm generates all of Company A's organizational functions:

- Marketplace positioning that is fixed;

- Organizational alignment that is politically negotiated with customer organizations as well as executives;

- Human performance that is directed by S-O-R discriminative learning or branching systems;

- Information modeling that is driven by S-R conditioned responding systems;

- Mechanical tooling that is implemented by s-r mechanical chaining systems.

Company B, in turn, is process-centric within the possibilities paradigm: the engineers process for and with the phenomena of the product in order to generate their most powerful images. As we can see below, in Figure 5-17, the processing operations are focused upon the phenomena of the product. Moreover, because the images are continuously changing, the processing systems are substantive, not political: they emphasize the substance of the operations in a never-ending attempt to generate more powerful images of the product.

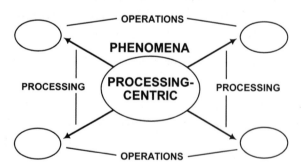

**Figure 5-17. The Process-Centric
Possibilities Paradigm**

This process-centric paradigm generates all of Company B's organizational functions:

- Marketplace positioning that is continuously evolving and substantively driven by product-engineering principles;

- Organizational alignment that is continuously generated by concurrent process engineering;

- Human processing that is empowered to generate by S-P-R generative processing systems;

- Information modeling that is driven by S-O-R discriminative learning systems;

- Mechanical tooling that is implemented by S-R mechanical processing systems.

Indeed, companies A and B contrast vividly in the processing systems available to them to discharge organizational functions. Table 5-5, on the following page, illustrates this contrast:

- Company A does not even have processing systems for discharging organizational functions; it lacks…

    - S-MP-R marketplace processing systems,
    - S-OP-R organizational processing systems,
    - S-P-R human processing systems.

- The remainder of Company A's systems are slotted one level behind Company B's elevated processing systems.

As a consequence, Company A brings the conditioning paradigm of the Industrial Age to the marketplace. After the goals are set and the plans are drawn, the entire assembly systems decompose into conditioned responding systems:

- Fixed marketplace positioning,
- Politically negotiated organizational alignment,
- S-O-R discriminative planning systems,

- S-R conditioned modeling systems,
- s-r conditioned-chaining tooling systems.

Table 5-5. Comparative Processing Systems

	COMPANY A	COMPANY B
MARKETPLACE POSITIONING	Fixed Marketplace Positioning	S-MP-R Marketplace Processing
ORGANIZATIONAL ALIGNMENT	Politically Negotiated Organizational Alignment	S-OP-R Organizational Processing
HUMAN PROCESSING	S-O-R Discriminative Learning	S-P-R Generative Processing
INFORMATION MODELING	S-R Conditioned Responding	S-O-R Discriminative Learning
MECHANICAL TOOLING	s-r Conditioning Chains	S-R Conditioned Responding

In turn, Company B emphasizes the evolving process-centric requirements of the Information Age:

- S-MP-R marketplace positioning systems,
- S-OP-R organizational alignment systems,
- S-P-R human processing systems,
- S-O-R discriminative modeling systems,
- S-R conditioned-responding tooling systems.

In short, Company B's process-centric hybrid paradigm generates new capital development, as shown next.

S-MP-R ⟶	**MCD, or marketplace capital development**
S-OP-R ⟶	**OCD, or organizational capital development**
S-P-R ⟶	**HCD, or human capital development**
S-O-R ⟶	**ICD, or information capital development**
S-R ⟶	**mCD, or mechanical capital development**

In summary:

• The planning paradigm *"freezes"* the design of the product; consequently, the company is stuck with the basic design it began with.

• The processing paradigm *"frees"* the product to evolve in its most functional form; thus, the company's future leaders will have a spiraling array of *"virtual"* product designs from which to draw for their specific purposes.

Which one does investment capital track in our newly integrated and interdependent global economy?

IN TRANSITION

The future of civilization lies with the new Monolithic Idea. Before its implementation, movement is random and sporadic. Crises creates "regression to the mean" as marginal members of society grope inductively and quantitatively for passage to the next level of development. After its implementation, movement is continuous and systematic as all members contribute their processing interdependently for larger, common purposes. Not only thinking people are freed to contribute, but also thinking organizations and thinking communities.

159

Historically, monolithic ideas have integrated previously disparate entities and, in so doing, have introduced civilization to an elevated plane of functioning. Thus, just as the historic monolithic ideas projected us from hunter-gatherers to herder-farmers and created the very notion of civilization, so will the new M.I. catapult us through the integration of capital resource and processing development.

In this context, the historic functions of science—to describe, predict, and control variability—give way to the futuristic function of science: to free changeability through empowering explosions of processing in the phenomena to which we are relating. We may represent such processing with an interdependent-generativity-intelligence illustration of our NCD systems, as presented in Figure 5-18. As may be noted, the I^5 interdependent processing systems rotate around the NCD systems: MCD, OCD, HCD, ICD, and mCD. We may compare this figure to the illustrations of inductive, deductive, and generative intelligence with NCD systems, presented earlier (Figures 2-26, 3-26, and 4-16, respectively). The multiple, *nested* processing systems are exponentially more powerful in generating phenomenal information.

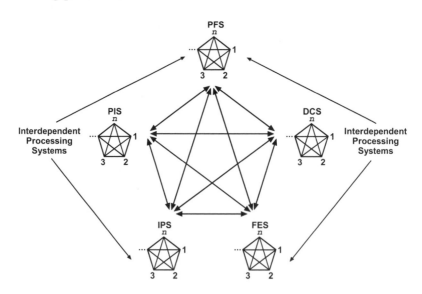

Figure 5-18. Interdependent-Generativity-Intelligence Illustration of NCD

From the broadest perspective, the freeing of human contributions yields the freeing of human destiny in time and space. Time is created through the systematic integration and application of processing systems. Space is created through the penetration of processing systems in internal space—within marketplaces, organizations, humans, information, and machinery.

In transition, the central theme of human history has been to employ human intelligence to free new sources of variability in all areas of human endeavor—scientifically, educationally, economically, politically, socially. The new M.I. generates our human destiny in the image of our own infinitely powerful intelligence and, in so doing, empowers us to become truly human.

6 Possibilities Processing and Paradigmatic Modeling

Generative processing may emanate from any of the cells in the possibilities science model (see Figure 6-1). As shown below, directional movement is represented by the interdependent processing symbols, with generative initiatives issuing from all of the source cells. This means that the possibilities processor may start with any cell and move in any direction.

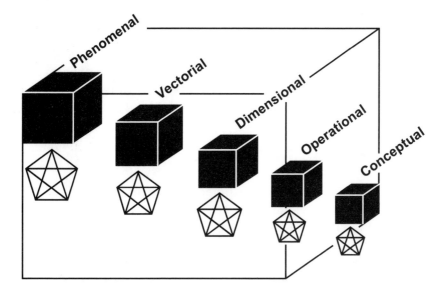

Figure 6-1. Possibilities Processing

Again, dimensional intelligence is the prerequisite for generative intelligence. With dimensional intelligence—whether deductively derived or developmentally induced—generative processing is possible. Without dimensional intelligence, generativity is not possible. With both deductively derived and developmentally induced dimensional intelligence, the highest levels of generative processing are possible.

Remember, dimensional intelligence simply means that we have defined the phenomena—whatever phenomena concern us—multidimensionally: their functions, components, processes; their conditions and standards. Dimensional intelligence allows us to relate all of these dimensions interdependently. Moreover, dimensional intelligence empowers us to generate new and more powerful dimensions and, thus, new and more powerful phenomena. The generation of these new and more powerful dimensions is the core of generative processing.

We have already learned that, under most circumstances, we initiate processing with inductive or developmental modeling; in turn, we culminate processing with generative processing. In the absence of deductive models, we begin processing inductively with the fragments of data available to us: we look for patterns or conceptual relationships; we hypothesize the operations of these relationships; we formulate dimensional models for these operations. The dimensional models that we generate will guide our deductive processing to hypothesis-testing.

Paradigmatic Modeling

When well-defined deductive models and formulas are available, we can initiate deductive processing by deriving vectorial and dimensional models and testing the hypotheses they generate. The results of the hypothesis-testing enable us to innovate on the models that we or others have generated.

By way of example, we may begin with the **Cultural Capital Development (CCD) Matrix,** shown in Table 6-1. As we can see, the CCD functions emphasize relating dependently, competitively, independently, collaboratively, and interdependently. This is the essence of CCD: healthy cultures relate collaboratively and interdependently; unhealthy cultures relate dependently or not at all because of their conditioning.

As may be noted, these relating functions are discharged by economic components ranging from free-enterprise economics to control economies. Finally, governance processes enable the economic components to discharge the relating functions; those processes range from free and direct democratic governance to totalitarian systems. Their position on the diagonal means that they represent *"the z dimension,"* or third dimension, of processing.

Table 6-1. Multiparadigmatic Modeling— CCD Matrix

ECONOMIC COMPONENTS

RELATING FUNCTIONS	Free Enterprise	Capitalistic	Mixed	Command	Control
Interdependent	Free Democratic				
Collaborative		Representative			
Independent			Mixed		
Competitive				Authoritarian	
Dependent					Totalitarian

We may summarize the generic mission for CCD as follows:

Relating functions are discharged by economic components enabled by governance processes.

Of course, at the highest level of CCD to which we may aspire, we may summarize the CCD mission in more idealistic terms:

Interdependent relating functions are discharged by free-enterprise economic components enabled by free and direct democratic governance processes.

This is the mission for CCD in the global marketplace of the twenty-first century. Whoever achieves this mission elevates our global civilization.

When we rotate the CCD Matrix deductively, we develop the **Economic Matrix** presented in Table 6-2. Here, our economic functions are discharged by our governance components. As may be noted, our new capital development (NCD) systems are processes that enable governance components to accomplish economic functions. We may summarize the mission of the Economic Matrix operationally:

> *Economic functions are discharged by governance components enabled by NCD processes.*

This is the economic mission that yields the largesse to which everyone everywhere aspires. The enabling NCD processing systems are the focus of the illustration of multiparadigmatic organizational modeling discussed next.

Table 6-2. Multiparadigmatic Modeling— Economic Matrix

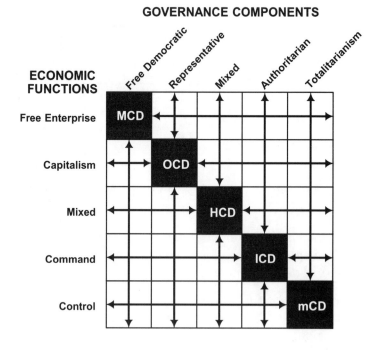

We may view multiparadigmatic modeling in depth by rotating deductively to the **MCD Matrix** (see Table 6-3). Here, marketplace requirements are discharged by corporate technologies: marketplace technologies (MT), organizational technologies (OT), human technologies (HT), information technologies (IT), and mechanical technologies (mT). Note that the *source cells* of the enabling organizational alignment systems are represented diagonally: leadership, marketing, resources (and their integration), technology, production. Again, this diagonal represents *"the z dimension"* in multidimensional modeling.

**Table 6-3. Leadership-Driven Organizational
Systems in MCD Matrix**

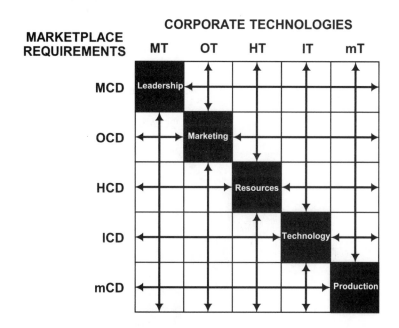

We may define all of the organizational alignment *source cells* within the MCD Matrix in operational terms:

- *MCD-positioning requirements are discharged by marketplace technologies enabled by leadership systems.*

- *OCD-alignment requirements are discharged by organizational technologies enabled by marketing systems.*

169

- *HCD-processing requirements are discharged by human technologies enabled by resource integration systems.*

- *ICD-modeling requirements are discharged by information technologies enabled by technology development systems.*

- *mCD-tooling requirements are discharged by mechanical technologies enabled by production systems.*

We may also define any of the cells in our matrix operationally. The MCD Matrix can be summarized operationally as follows:

> *MCD requirements are discharged by corporate technologies enabled by organizational alignment systems.*

Similarly, we may view multiparadigmatic modeling in the **OCD Matrix,** presented in Table 6-4. As may be noted, MCD

Table 6-4. Goaling-Driven Human Processing Systems in OCD Matrix

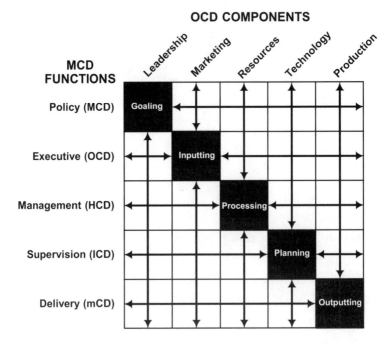

functions are discharged by OCD components. As may also be noted, the *source cells* of the enabling human processing systems are represented diagonally: goaling, inputting, processing, planning, outputting.

We may define all of the human processing *source cells* within the OCD Matrix in operational terms:

- *Policy-positioning functions are discharged by leadership components enabled by goaling systems.*

- *Executive-alignment functions are discharged by marketing components enabled by inputting systems.*

- *Management-processing functions are discharged by resource-integration components enabled by human processing systems.*

- *Supervisory modeling functions are discharged by technology-development components enabled by planning systems.*

- *Delivery-tooling functions are discharged by production components enabled by outputting systems.*

Again, any of the cells in our matrix can be defined operationally.

We may summarize the OCD Matrix operationally in the following way:

> *MCD functions are discharged by OCD systems enabled by human processing systems.*

We may also view multiparadigmatic modeling in the **HCD Matrix,** shown in Table 6-5. As we can see, OCD functions are discharged by HCD components. Again, the *source cells* of the enabling systems—this time, information modeling—are represented diagonally: phenomenal, vectorial, dimensional, operational, conceptual.

Table 6-5. Phenomenal-Driven Information Modeling Systems in HCD Matrix

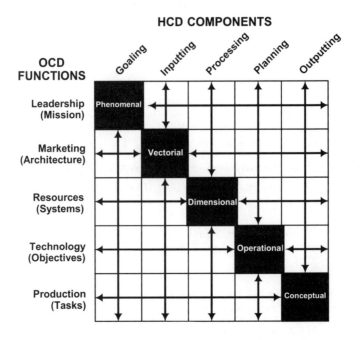

We may summarize all of the information-modeling *source cells* within the HCD Matrix in operational terms:

- *Leadership-missioning functions are discharged by goaling components enabled by phenomenal modeling systems.*

- *Marketing-architectual functions are discharged by inputting components enabled by vectorial modeling systems.*

- *Resource-systems-integration functions are discharged by processing components enabled by dimensional modeling systems.*

- *Technology-development-objective functions are discharged by planning components enabled by operational modeling systems.*

- *Production-task-performance functions are discharged by outputting components enabled by conceptual modeling systems.*

172

Again, any of the cells in our matrix can be defined operationally. We may summarize the HCD Matrix operationally in this way:

OCD functions are discharged by HCD components enabled by information-modeling systems.

Likewise, we may view multiparadigmatic modeling in the **ICD Matrix** (see Table 6-6). Here, HCD functions are discharged by ICD components. Again, the *source cells* of the enabling systems— this time, mechanical tooling—are represented diagonally: functions, components, processes, conditions, standards.

Table 6-6. Functions-Driven Mechanical Systems in ICD Matrix

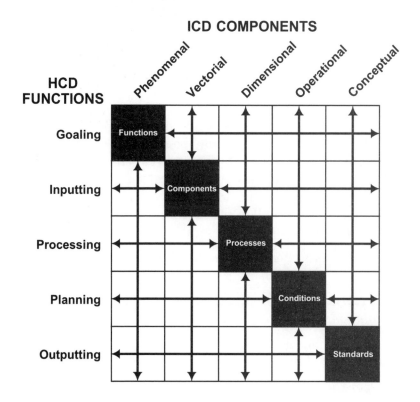

We may summarize all of the mechanical-tooling *source cells* within the ICD Matrix in operational terms:

- *Goaling functions are discharged by phenomenal components enabled by functionalizing systems.*

- *Inputting functions are discharged by vectorial components enabled by "componentizing" systems.*

- *Processing functions are discharged by dimensional components enabled by processing systems.*

- *Planning functions are discharged by operational components enabled by conditioning systems.*

- *Outputting functions are discharged by conceptual components enabled by standardizing systems.*

Again, any of the cells in our matrix can be defined operationally. We may summarize the ICD Matrix operationally as follows:

HCD functions are discharged by ICD components enabled by mCD tooling systems.

Finally, we may view multiparadigmatic modeling in the **mCD Matrix,** shown in Table 6-7. As may be noted, ICD functions are discharged by mCD components. The *source cells* of the enabling mechanical-programming processes are represented diagonally: design, direction, tasks, steps, implementation.

We may summarize all of the mechanical programming *source cells* within the mCD Matrix in operational terms:

- *Phenomenal modeling functions are discharged by functional components enabled by designing processes.*

- *Vectorial modeling functions are discharged by operational components enabled by directing processes.*

- *Dimensional modeling functions are discharged by operational processes enabled by tasking processes.*

- *Operational modeling functions are discharged by conditioning components enabled by step-by-step programs.*

Table 6-7. Design-Driven Mechanical Programs in mCD Matrix

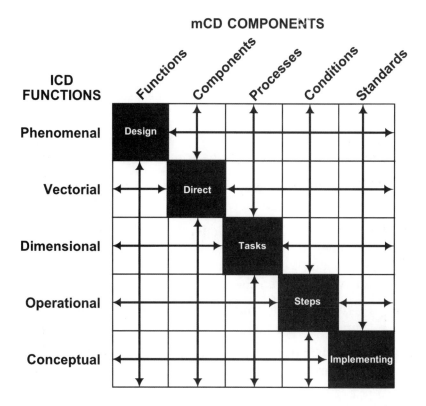

* *Conceptual modeling functions are discharged by standardizing components enabled by implementation programs.*

We may also define any of the cells in our matrix operationally. Finally, we may summarize the mCD Matrix operationally:

ICD functions are discharged by mCD components enabled by mechanical programming systems.

To summarize, these are the multiparadigmatic tools of the NCD system: MCD, OCD, HCD, ICD, mCD. They are vectorial models that give direction to our modeling. They empower us to generate new organizational images and initiatives.

Possibilities Processing:
A Case Study

This section presents a case study of possibilities processing with our organization Human Technology, Inc. Two choices were made in preparation for this processing:

- First, we selected the function we wished to discharge. That function was to empower the organization for processing in the twenty-first-century global economy.

- Second, we selected our images of organizational modeling. Because we were addressing an operational organization, we selected operational information modeling.

We then employed our interdependent processing systems to enable the operational information to empower the phenomenal organization. We initiated processing inductively: I^1, I^2, I^3, I^4, I^5. Thus we employed the following:

- I^1—Information relating systems to define the organizational phenomena operationally;

- I^2—Information representing systems to model images of the organizational phenomena;

- I^3—Individual processing systems to generate new images of the organizational phenomena;

- I^4—Interpersonal processing systems to generate improved images of the organizational phenomena;

- I^5—Interdependent processing systems to generate prepotent images of the organizational phenomena.

In the above context, information modeling is the precondition for all human processing. Information modeling enables us to represent images of the phenomena we are addressing in our processing. Before we can model information, we must relate to information. Yes, we must relate to data as if it were alive, for it is indeed alive! We must transform this data into conceptual information that describes our images conceptually in terms of their relationships. Finally, we must transform our conceptual images into operational images that define the phenomena in terms of their operations. Again, operational definitions enable us to represent the information in multidimensional information models.

I^1—Information Relating Systems

As C.O.O. of Human Technology, Inc., Sharon Fisher understood the importance of the company's marketplace positioning (see Figure 6-2). She knew the company was positioned to maximize its comparative advantage in the marketplace. She also knew that such market-driven positioning required a flexible and *"super-responsive"* organization dedicated to responding to subtle changes in the marketplace; in short, that it required a *"possibilities organization"* aligned to implement marketplace positioning.

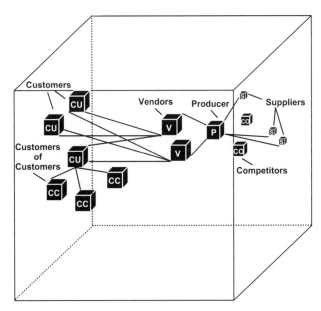

Figure 6-2. Marketplace Positioning

In this context, Fisher had an image of the operational definitions she needed to generate through relating. She defined the operations generically:

- **Functions,** or purposes or outputs;
- **Components,** or parts or inputs;
- **Processes,** or procedures or methods;
- **Conditions,** or contexts or environments;
- **Standards,** or levels of achievement or excellence.

Well aware that she also needed to know the organization well enough to tailor these operational definitions, Fisher related to the organizational experience in order to define its operations. Her initial operational definition is shown below, in Table 6-8.

Table 6-8. Operational Definition of Organization

COMPONENTS	Organizational units are "architected"
FUNCTIONS	to align with marketplace positioning
PROCESSES	by implementing human processing
CONDITIONS	in the context of marketplace requirements
STANDARDS	at uniform standards of human performance.

Our possibilities scientist was pleased with this definition: it recognized that organizations are built to align with marketplace positions; it also emphasized that organizational functions are implemented by human processing. However, she was concerned about the limitations of her definition for the organization. It did not take into consideration the continuously changing marketplace requirements; therefore, it did not define the organization's continuously changing operations. Accordingly, Fisher continued her relating, and penetrated to a deeper and more powerful level of definition (see Table 6-9).

Table 6-9. New Operational Definition of Organization

COMPONENTS	**Organizational units are continuously "architected"**
FUNCTIONS	**to align with continuous marketplace positioning**
PROCESSES	**by implementing continuous interdependent processing**
CONDITIONS	**in the context of spiraling marketplace requirements**
STANDARDS	**at increasingly diverse and changeable standards of human processing and information modeling.**

As can be seen, our possibilities scientist gave the operational definition a whole new meaning—intentionality! The organization is defined by continuous alignment with continuous changes in marketplace requirements and positioning; by continuous interdependent processing with increasingly diverse and changeable standards of performance.

Through her successful efforts in this relating phase, Fisher confirmed the following:

- *All things are possible with relating; nothing is possible without it.* Without relating, there are only probabilities—ultimately, probabilities for failure; with relating, there are potentially infinite possibilities.

  Specifically, the operational information generated by information relating systems enable information representing systems: together, they constitute preparation for processing, serving to introduce managers and phenomena to each other's possibilities.

- *The power of information relating relies in large part on attitude:* an attitude of humility, to be sure—a recognition that no one has all the answers and that all answers will be discovered by processing with the phenomena!

Most important, this possibilities scientist concluded that information relating signifies interdependency with our universe—our workstation, team, unit, organization, customers, suppliers, vendors, and marketplace. As we fully enter the twenty-first century, it is clear that no one and nothing are independent. Perhaps such independence was only a twentieth-century illusion!

I²—Information Representing Systems

Armed with an understanding of the organization's operations, Fisher entered the next phase in interdependent processing: building a dimensional image of the organization. She needed to "see" the organization; in particular, she needed to see how every dimension related to every other dimension. Accordingly, she began to "dimensionalize" the operations of the organization.

This possibilities scientist already had a good image of the organization's functions (see Table 6-10). She understood the functions as follows:

- **Policy** defined the missions to implement marketplace positioning;

- **Executives** designed the organizational architecture to align with marketplace positioning;

- **Management** designed the systems to achieve the goals of organizational alignment;

- **Supervision** defined the objectives to implement the systems;

- **Delivery** performed the tasks to achieve the objectives.

She now needed to define the resource components that accomplished these functions.

Table 6-10. Organizational Functions

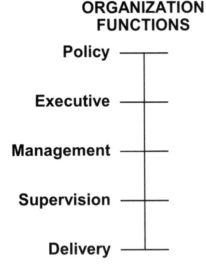

ORGANIZATION FUNCTIONS

Policy

Executive

Management

Supervision

Delivery

Our possibilities scientist initiated "dimensionalization" by developing the organizational components that discharged the functions. This development entailed constructing a two-dimensional matrix for defining the organizational components (see Table 6-11). In working with the matrix, Fisher came to understand the components as follows:

- **Leadership** defined the new marketplace directions to reflect marketplace positioning;

- **Marketing** developed new marketing relationships with customers to reflect marketplace directions;

- **Resource integration** developed tailored solutions for customers in new marketing relationships;

- **Technology** developed customized designs for products and services in new marketing relationships;

- **Production** produced standardized products and services in new marketing relationships.

She now needed to define the organizational processes that enabled the components to discharge the functions.

Table 6-11. Organizational Functions and Components

COMPONENTS

FUNCTIONS	Leadership	Marketing	Resources	Technology	Production
Policy					
Executive					
Management					
Supervision					
Delivery					

Fisher continued to dimensionalize by developing the organizational processes that enabled the components to discharge the functions (see Figure 6-3). She did this by developing a three-dimensional model that defined the organizational processes:

- **Goaling,** or measuring values;
- **Inputting,** or analyzing operations;
- **Processing,** or synthesizing new operations;
- **Planning,** or operationalizing new objectives;
- **Outputting,** or technologizing new programs.

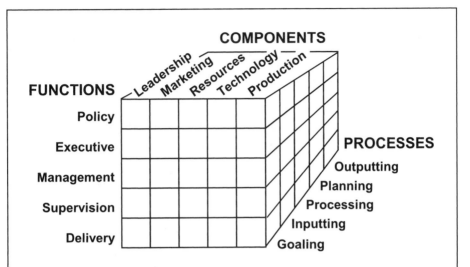

**Figure 6-3. Organizational Functions,
Components, and Processes**

Fisher had now developed a three-dimensional model, and could "see" how every level of every dimension related to every other dimension. This gave her a great perspective on organizational operations. However, she had not dimensionalized all of the operations.

Paramount among the operations were the conditions within which the organization operated (see Figure 6-4). These conditions, she found, generated the functions of the organization. As such, they were prepotent factors in organizational functioning. For the time being, she was satisfied to identify the conditions in terms of the marketplace. She noted that the marketplace conditions were defined by their own functions, components, and processes.

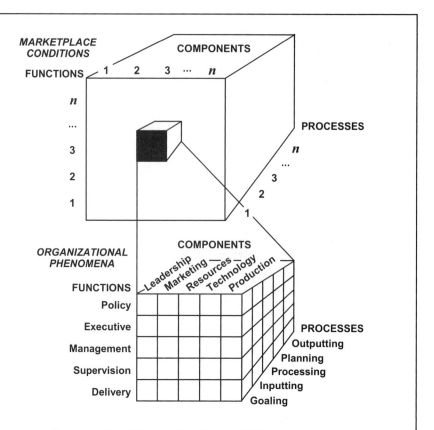

**Figure 6-4. Organizational Phenomena
and Marketplace Conditions**

Finally, our possibilities scientist needed to dimensionalize standards in order to culminate organizational operations (see Figure 6-5). Fisher knew that we cannot measure organizational functioning until we have assessed standards; however, she also knew that she was defining standards in a very different manner. She noted that the standards were drawn from the organizational units and defined in terms of human performance within the units. Consequently, as shown in our illustration, human standards were defined by leadership functions discharged by comprehensive human processing enabled by information modeling.

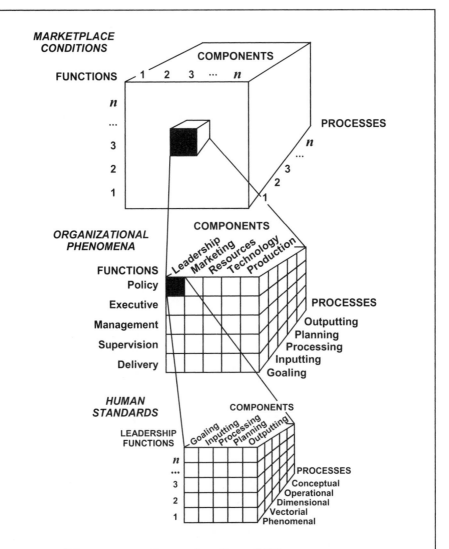

**Figure 6-5. Organizational Phenomena,
Conditions, and Human Standards**

Fisher had completed her information-representing assignment by dimensionalizing all of the organizational operations. She understood the power of the marketplace conditions in generating organizational functions. She also recognized the critical nature of the interactions of organizational dimensions, and realized that the standards of human performance define organizational effectiveness.

This possibilities scientist could now "see" the irony of the organizational system she was designing—that even as she dimensionalized organizational operations, the organization was assessing her standards of performance; that her performance would be assessed by human standards generated by organizational phenomena. She realized that the very process of dimensionalizing occurred within the policy cube: policy functions discharged by leadership components enabled by goaling processes. With great eagerness, Fisher began to recognize the full power of interdependent processing for developing a super-responsive, *"possibilities"* organization.

In summary, our possibilities scientist now understood the power of dimensionalizing:

 All phenomena have definable operations.

 All phenomenal dimensions are interdependently related.

Through her efforts in this representing phase, she had discovered the following:

- *All things can be processed once they are represented dimensionally.* Without dimensional information, there can be no systematic processing; with it, there are potentially infinite possibilities. Dimensionally represented information introduces managers and phenomena to one another's dimensional possibilities.

- *Information representing is not only a skill, but also an attitude:* an attitude of respect, commitment, and perseverance. Respect for the complexity of phenomena! Commitment to actualizing phenomenal potential! Perseverance in working toward the growth of both manager and phenomena!

Multidimensional information modeling communicates this respect because the multidimensional emphasizes interdependency. It is the interdependent processing of multidimensional systems that yields growth for all parties. Accuracy in information representing is a critical requirement for the possibilities manager in the twenty-first-century global marketplace.

In addition, Fisher had begun to comprehend her own power in generating standards for measuring the excellence of her own performance. She understood that, as human capital, she was *nested* within the multidimensional conditions of her organization, and that she herself constituted the conditions for the standards of her people's performance.

Empowered by her perspective of dimensional models of conditions, this possibilities scientist was now prepared to process generatively.

I^3—Individual Processing Systems

To initiate this phase of processing, Fisher established organizational goals—in this case, the corporate mission (see Figure 6-6). First, she defined the mission's dimensions:

- Public-sector targets,
- Consulting- and training-service goals,
- Leadership-driven organizational strategies.

She then summarized the mission in this operational definition:

Consulting- and training-service goals are accomplished with public-sector targets by way of leadership-driven organizational strategies.

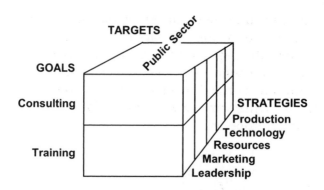

Figure 6-6. The Corporate Mission

Next, Fisher analyzed the current operations in terms of their ability to fill the super-responsive corporate mission (see Figure 6-7). She broke the operations down as follows:

Components inputs are transformed into functions outputs by processing.

Fisher assessed the standards and discovered that the current inputs were ineffective and inefficient in accomplishing the outputs under the current conditions. She decided to expand alternative operations.

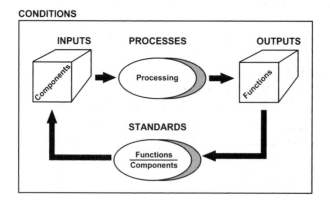

Figure 6-7. Analyzing the Current Corporation

Our scientist drew her images of alternative models from information representing systems (see Figure 6-8). She expanded her alternatives by considering all functions, components, and processes in these systems. The combinations and permutations of the dimensions at all of these levels would enable her to synthesize new organizational models to meet her corporate mission.

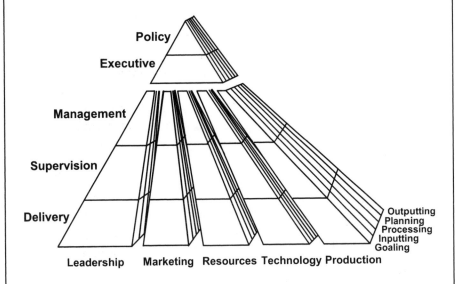

Figure 6-8. Synthesizing New Models

The illustration below (Figure 6-9) shows how Fisher operationalized her newly synthesized model. As may be noted, her organizational model is centralized by mission and decentralized by operations. In this image, the policy and executive levels have the responsibility for leading and "architecting" the organization. In turn, the marketing, resource integration, technology, and production operations are decentralized. This decentralized design manages the information problem and moves the corporation toward becoming super-responsive to the marketplace.

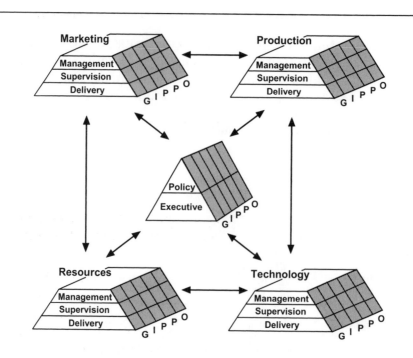

Figure 6-9. Operationalizing New Models

In technologizing the new model, Fisher found (among many nuances) that teams were more productive than formal hierarchical levels (see Figure 6-10). Accordingly, she designed management-led, team-based operations, thereby eliminating one level of hierarchy. With input and feedback from the people involved, she designed an increasingly flexible and responsive striking force, thereby further alleviating the oppressive effects of the monolithic organization.

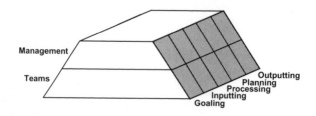

Figure 6-10. Technologizing the New Model

Our possibilities scientist was now approaching completion of her assignment. Because of her individual processing skills, she already had an early image of a super-responsive organizational model, including:

 Operational phenomena,

Dimensional phenomena,

Generative phenomena.

It remained for her to engage others in interpersonal processing to generate the most powerful images of super-responsiveness.

Through her efforts in this phase, Fisher realized that her ability to process generatively was contingent upon her ability to model dimensionally. Only with skills in information representing could she expand the operations alternatives. Only with skills in information representing could she narrow the alternatives to the most highly leveraged operations. She concluded that generativity in thinking was a function of dimensionality in modeling.

In short, Fisher found that processing is prepotent among all skills: it is all-powerful in generating new ways of doing new things. She never felt "boxed-in" by any problems, and now there are no situations in which she feels she cannot generate "a better idea." That is because she has "a better process." Indeed, processing is simply more potent than the problems themselves. To be sure, processing is what makes people feel potent!

I^4—Interpersonal Processing Systems

Equipped with the phenomenal images generated by individual processing, Fisher was prepared for interpersonal processing. She related with co-workers "up, down, and sideways," employing her interpersonal processing systems:

- **Goaling** by measuring values,
- **Getting** others' images of the organization,
- **Growing** new images of the organization,
- **Going** on to implement the new images.

As a result, the group of co-workers expanded the initial corporate mission (see Figure 6-11), altering it to incorporate the following:

- Private-sector targets,
- Software-product and -services goals,
- Organizational strategies driven by research and development (R and D).

Together, Fisher and the others summarized the newly redefined mission:

> *Consulting, training, and software goals are accomplished with private- and public-sector targets by way of R-and-D-driven strategies.*

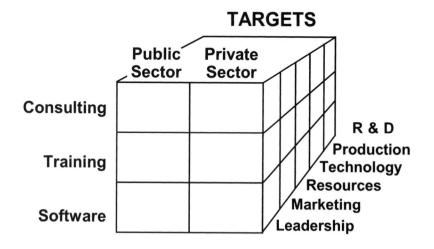

Figure 6-11. The Expanded Mission

Our possibilities scientist continued to employ her interpersonal processing systems **to get** the others' images of the organization (see Figure 6-12). Here it is important to understand that the co-workers had engaged in individual processing on their own before entering interpersonal processing. One of their major insights was to invert the organization. This placed the delivery people at the source of ongoing information bulletins. They believed this was most appropriate in a service industry.

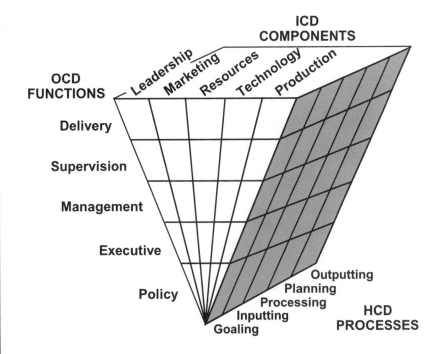

Figure 6-12. Getting Organizational Images

Next, Fisher used her interpersonal processing skills **to give** her own image of organizational operations (see Figure 6-13). Here she shared with the group the image that she had generated in individual processing, with the organization centralized by mission and decentralized by operations.

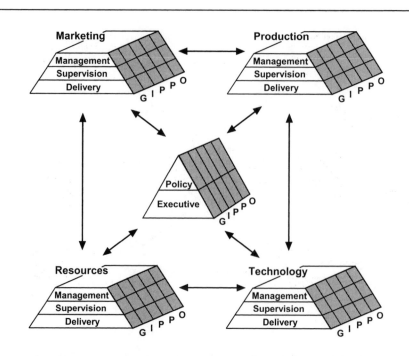

Figure 6-13. Giving Organizational Images

She and the group then employed their interpersonal processing skills **to grow** a new image of organizational operations (see Figure 6-14). This was the major work of interpersonal processing: using various images as input to generate entirely new images of organizational operations. We can describe those operations as:

- Centralized by mission,
- Decentralized by operations,
- Complemented by partnerships,
- Inverted where appropriate,
- Teamed where appropriate.

Together, our interpersonal processors had generated a possibilities organization: an organization that could instantaneously configure its functions, components, and processes to meet any changing market conditions and human standards.

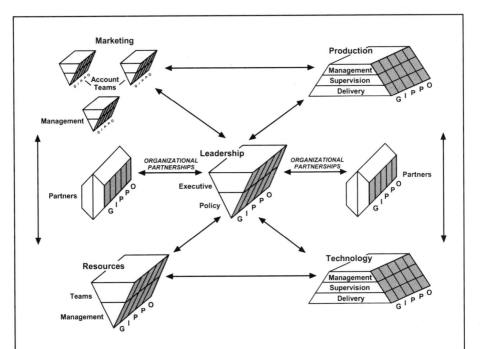

Figure 6-14. Growing New Organizational Images

The group employed interpersonal processing skills **to go** on to plan implementation of the new organizational model (see Figure 6-15). Members processed, among technological nuances, entities such as inverted teaming. Here teams are in the lead positions to receive information, to process it, and to initiate. In turn, management is cast in a support role for higher-order processing.

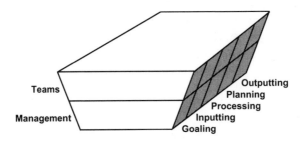

Figure 6-15. Going on to Technologize New Images

At this point in processing, the group was ready to implement a super-responsive organizational model that included:

I^1 *Operational phenomena,*

I^2 *Dimensional phenomena,*

I^3 *Generative phenomena,*

I^4 *Improved generative phenomena.*

The I^4 phase led to many discoveries for Fisher and her group. One of the most exciting was that interpersonal processing applications can take a multitude of forms, depending on the diversity of the interpersonal processors and their skills. For instance, further generative interpersonal processing generated images of possibilities organizations such as the one presented in Figure 6-16. A possibilities organization is capable of instantaneously aligning resources to serve continuously changing policy.

We may note the following about our illustration:

- The policymaking function and the leadership component are decentralized to provide maximum horizontal and verbal integration in leadership.

- Similarly, the technology and production components are decentralized to maximize the integration of information modeling and mechanical tooling.

- Finally, the marketing components and delivery functions are decentralized to address special needs.

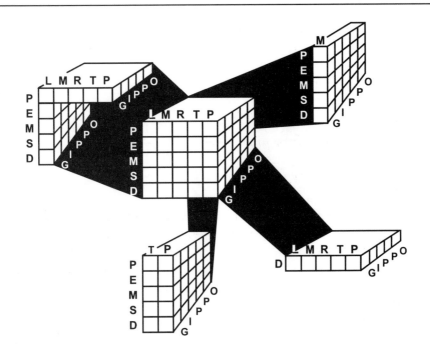

Figure 6-16. Images of a Possibilities Organization

I⁵—Interdependent Processing Systems

Empowered by the image of "the possibilities organization," Fisher and her colleagues were able to *nest* the organization inside marketplace conditions, as shown in Figure 6-17. Note that the possibilities organization was designed to be aligned with marketplace positioning. Fisher and her group then prepared to *nest* inside the possibilities organization all those requirements for generating organizational capital: human processing, information modeling, and mechanical tooling.

This means that each level of new capital development was contingent upon every other level. Clearly, marketplace positioning required organizational alignment to become marketplace capital. That was the mission of Fisher and company's interdependent-processing effort. Similarly, "the possibilities

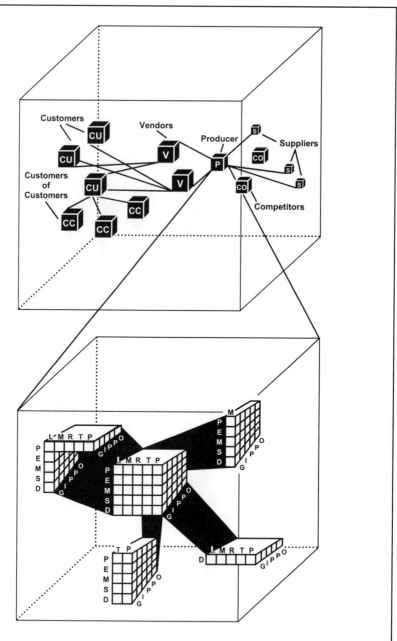

Figure 6-17. The Possibilities Organization
Nested **Within Marketplace Positioning**

organization" could be converted to organizational capital only with successful human processing. Likewise, human processing could become human capital only with effective information modeling. Finally, information modeling could become information capital only with productive mechanical tooling.

Fisher and her group realized that these forms of new capital development are not static, but continuously changing, and so managers are continuously processing. They also realized the implications of this for the managers in their organization: now that those managers had helped to generate "the possibilities organization" design, they had to learn to live and work within that design in order to generate organizational capital. Above all, they had to engage in continuous interdependent processing—with one another and with the phenomena.

With this in mind, the group worked on the concept of processing teams, representing them by the interpersonal processing model shown in Figure 6-18. We can see that the individual processors process independently before coming together to process interpersonally. Note that the teams can generate at least four generations of responses (R^4) in a relatively brief period of time.

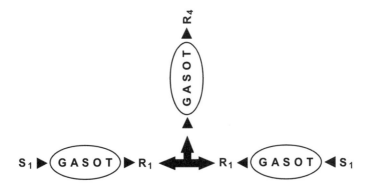

Figure 6-18. Processing Teams

Next, these interpersonal processing teams were *nested* within the phenomenal systems (see Figure 6-19). The mission was continuous interdependent processing with the phenomenal systems; thus, the interpersonal processing teams would bring their leveraged potential to bear upon the interdependent processing mission. In this manner, the group generated a design for organization capital development.

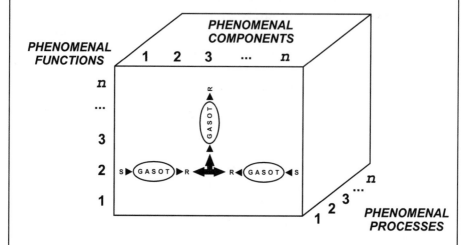

Figure 6-19. Interdependent Processing Teams
***Nested* Within Phenomenal Systems**

Fisher and company now addressed the issue of human capital. They realized that human processing required information modeling to become such capital. Accordingly, Fisher introduced NCD-systems modeling to increase the potential alternative images of the interdependent processing teams. We can see this modeling in Figure 6-20. Notice that the higher-order levels of new capital development *nest* the lower-order levels, from MCD to mCD.

- MCD—Marketplace capital development, or continuous marketplace positioning;

- OCD—Organizational capital development, or continuous organizational alignment;

- HCD—Human capital development, or continuous human processing;

- ICD—Information capital development, or continuous information modeling;

- mCD—Mechanical capital development, or continuous mechanical tooling.

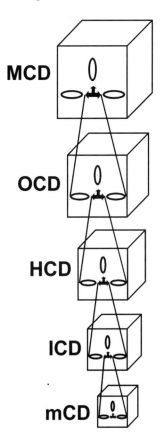

Figure 6-20. Interpersonal Processing Within New Capital Development Systems

This related all capital development systems interdependently. Moreover, in the interdependent and synergistic processing between HCD and ICD, people would have to learn to process within the information models they had developed.

In short, Fisher and her group began to develop a phenomenal perspective of the possible:

I^1 *Operational phenomena,*

I^2 *Dimensional phenomena,*

I^3 *Generative phenomena,*

I^4 *Improved generative phenomena,*

I^5 *Super-generative phenomena.*

The group knew it needed to prepare for the interdependent processing of these interdependently related organizational phenomena.

Again, Fisher and her colleagues were excited about their discoveries. They found that interdependent processing applications also can take many unique forms, depending upon the unique interactions of the phenomenal processing with the interpersonal processing. For instance, further generative interdependent processing resulted in images of partnered possibilities organizations such as those shown in Figure 6-21. Here the partnered organizations share both marketing components and delivery functions. Because they address the same marketplace with the same issues, the partners process interdependently within the marketing component. Because the partners also have the same delivery functions, they process interdependently to deliver products and services.

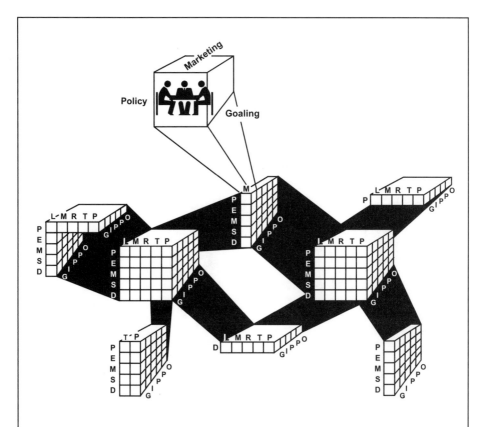

Figure 6-21. Partnered Possibilities Organizations

As detailed in Figure 6-22, the interdependent processing teams are *nested* within each of the units. In our illustration, the partnered processing teams are engaged in interdependent processing within the following cell: *marketing components discharging policymaking functions through goaling*. As may be noted, the processing teams are processing for the organizational unit by considering marketplace requirements and organizational capacity. The teams process interdependently with this positioning matrix to position the company to meet ICD and mCD requirements with IT and mT corporate capacities.

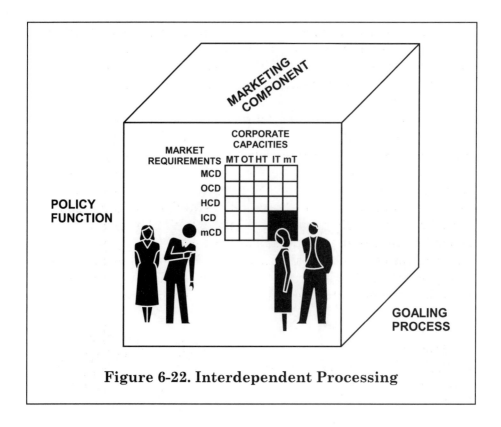

Figure 6-22. Interdependent Processing

In summary, the purpose of interdependent processing is to relate interdependently to phenomena, their conditions, and their standards of performance. As possibilities scientists put it, *we become **one** with the phenomena, but more than that, we become **multiple** with the changing phenomenal experiences.*

Everything is interdependently related to everything else, and all things—including ourselves—are changeable. This is the basic principle of interdependent processing, that ***everything is interdependent and changeable***. Are we willing to elevate our processes to relate to our changing, interdependent world? In contemplating this question we may remember that possibilities processing incorporates all of the processing systems already presented: inductive, deductive, generative, interdependent.

In review, we prepare for interdependent processing by relating to phenomena and representing them. We initiate interdependent processing by processing individually within the phenomena. However, we do not culminate interdependent processing until we become the phenomena themselves and process with their systems. Again, *we live in the house we built! And we keep changing it!*

Interdependent processing is the culminating stage in possibilities processing. After representing phenomenal images, the possibilities processor must develop images of the continuously changing conditions within which the phenomena operate, as well as images of the continuously changing standards that are generated within the phenomena (see Figure 6-23). The goal of interdependent processing, then, is to develop continuously changing perspectives of phenomenal possibilities. We do this by bringing our interpersonal and individual processing systems into interaction with the phenomenal systems that we have generated.

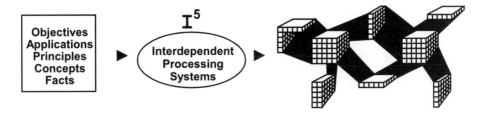

Figure 6-23. Interdependent Processing Systems

When we view these images, we realize the awesome power of possibilities processing. It transforms conceptual images of phenomena into powerful vectorial images of phenomena: in this instance, it transforms conceptual images of organizations into partnered possibilities organizations. That is the nature of possibilities science: to make something out of nothing but our processing abilities.

In transition, science is perspective and attitude—the perspective we gain from the processing experience; the attitude we gain from committing ourselves to the wisdom of processing. In our search for meaning in the universe, we are dedicated to these propositions.

The spiraling changes of the twenty-first-century global community and marketplace require a new and powerful human processor. The possibilities scientist is that processor. The possibilities scientist (PS) is empowered by all of our powerful and interrelated processing systems: scientific, interdependent, substantive (see Figure 6-24).

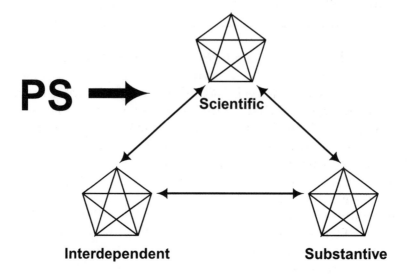

Figure 6-24. The Possibilities Scientist

We may symbolize these processing systems and the information that they generate by information capital development. In this context, the possibilities scientist (PS) is a product of the synergistic relationship between human capital development (HCD) and information capital development (ICD), as represented in Figure 6-25.

Figure 6-25. The Possibilities Scientist

To complete the formula, we must add the "A" factor: a free and changing ATTITUDE of commitment to processing, and the free and changing perspective it generates (see Figure 6-26).

$$PS \longrightarrow (HCD \longleftrightarrow ICD) + \text{"A"}$$

Figure 6-26. The Possibilities Scientist

A precondition of this ATTITUDE is knowing that descriptive data is postdictive of the past, not predictive of the future. A condition of this ATTITUDE is realizing that possibilities begin by our relating to the phenomena themselves. The power of this ATTITUDE is projecting the potential of phenomena based upon empowering phenomena. The culmination of this ATTITUDE is releasing the empowered phenomena, with all of their contingencies and vulnerabilities, to seek their own changeable destinies. Yes, the ATTITUDE culminates with freedom for the phenomena *and ourselves*, the possibilities scientists. In the final analysis, science is the perspective of processing and the ATTITUDE of freedom, for freedom is change!

III

Summary and Transition

The fundamental principles of possibilities science: phenomenal processing, modeling, and changing phenomena. The corollaries of possibilities principles: nesting, rotating, and transmitting genetic coding.

7 Concurrency and Paradigmetric Technologies

Thus far in our book, we have described a variety of projects to illustrate the new science of possibilities (see Figure 7-1). All of the projects are related in both operations and applications. All of the projects incorporate the basic operations of possibilities science, which are distinguished by the following features:

- They are dedicated to relating, empowering, and freeing phenomenal functions;

- They emphasize phenomenally driven information-modeling components;

- They are enabled by interdependent phenomenal processing systems.

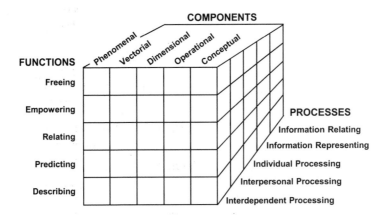

Figure 7-1. The New Science of Possibilities

These operations were derived from changeable conditions, which free possibilities science. They generate standards of changeability that serve to conduct continuing measures of performance. Their role is clear in this summary of the possibilities science mission:

> *Freeing functions are discharged by information-modeling components enabled by interdependent processing systems.*

In short, these operations define possibilities science. They make possible all initiatives—probabilities and possibilities!

Historic probabilities science found its contributions in content-centricity or specificity: the linear, independent, symmetrical, and static. Current probabilities science finds its contributions in network-centricity: the ultimate distribution of the content discovered at its "nodes." Indeed, its adherents believe in the "truth" yielded by the Internet and other modes of distribution with the willing participation of content-centric scientists.

In contrast, possibilities science is futuristic, continuously discovering its changeable "truths" in process-centricity. All phenomenal dimensions are processing dimensions: functions, components, processes, conditions, standards. All processing is continuous within, between, and among all phenomenal dimensions: multidimensional, interdependent, asymmetrical, and changeable. At the highest levels of all phenomenal dimensions, phenomena are to seek their own changeable destinies.

As we have seen, all of the possibilities projects illustrated in this book have the quality of interdependent relatedness; thus, they have meaning. Demonstrations in the marketplace and the community, in business and organizations, with education and individuals, all derive from the same basic model of possibilities science. Moreover, these diverse projects are, themselves, held together by the changeable conditions of process-centricity.

A deeper look into the illustrations will reveal the organizational model dedicated to new capital development (NCD) functions (see Figure 7-2):

 Marketplace positioning,

 Organizational alignment,

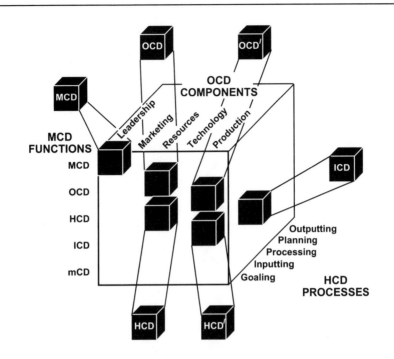

Figure 7-2. NCD Projects in Organizational Model

 Human processing,

 Information modeling,

 Mechanical tooling.

All of the projects illustrated may be mapped into the organizational functions, components, and processes. For example:

 Corporate productivity upgrade as dedicated to MCD (Chapter 4):

Marketplace-positioning functions are discharged by leadership-driven components enabled by goaling processes.

Hybrid organizational modeling as dedicated to OCD (Chapter 5):

> *Organizational alignment functions are discharged by marketing-driven components enabled by inputting processes.*

 Internal organizational-alignment modeling as dedicated to OCD′ (Chapter 3):

> *Internal organizational-alignment functions are discharged by resource-driven components enabled by generative processing.*

 Internal team- and individual-performance alignment modeling as dedicated to HCD (Chapter 2):

> *Internal human-processing alignment functions are discharged by marketing-driven components enabled by inputting processes.*

 Possibilities processing systems as dedicated to HCD′ (Chapter 6):

> *Human processing functions are discharged by resource-driven components enabled by generative processes.*

 Concurrent processing systems as dedicated to ICD (Chapter 1):

> *Information-modeling functions are discharged by resource-driven components enabled by generative processes.*

We may summarize the overall NCD mission as thus:

> *MCD functions are discharged by OCD components enabled by HCD processes.*

Again, our project illustrations have demonstrated the contributions of possibilities science to cultural capital development, or CCD (see Figure 7-3).

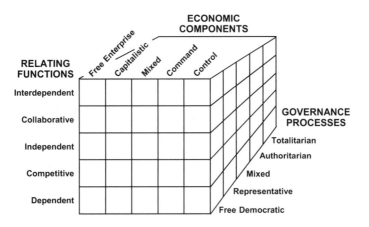

Figure 7-3. The CCD Model

All of the projects revolve around critical dimensions of the CCD model:

- Interdependent relating functions,
- Free-enterprise economic components,
- Free and direct governance processes,
- The enabling new capital development systems.

We may summarize the CCD mission as follows:

> *Interdependent relating functions are discharged by free-enterprise economic components enabled by free and direct governance processes.*

Together, these relating, economic, and governance systems define the CCD model for participation in the twenty-first-century global marketplace. These operations define the possibilities for culture change—the overriding issue of our future, if we are to have one! All of these dimensions are empowered by the new capital development systems.

Once again, the implications of new capital development for CCD are critical. NCD systems are the empowering capital development processes for all CCD. This is the organization's capital contribution to enabling governance and economic systems to discharge interdependent relating functions: MCD, OCD, HCD, ICD, mCD.

In this respect, science is the source of all cultural change. And possibilities science is the source of all cultural capital development:

- Freeing phenomenal functions defined by generative processing systems;

- Phenomenal components defined by information-modeling systems;

- Phenomenal processing systems defined by interdependent processing systems;

- Phenomenal conditions defined by freeing cultural contexts;

- Phenomenal standards defined by performance changeability.

In essence, possibilities science has an interdependent and synergistic relationship with cultural capital development: each grows with the other's contributions.

HCD ←⟶ ICD

The greatest challenge of our time or, indeed, any time is to deal with our information environments: to discriminate them, to develop them, to process them, to communicate them. They are not always discernable: we must define them operationally. They are seldom represented schematically; so we must model them dimensionally. They are rarely processed; so we must process them generatively. That is the challenge of education. That is the challenge of civilization. That is the challenge of becoming human.

In this context, managers merge their identities with their organizational functions; teachers with their contents; parents with their children. All view the world through the phenomenal experience of the people, data, and things of their existence. That is what interdependent processing is all about: generating new and powerful images of phenomenal possibilities; innovating new and powerful images within the phenomenal possibilities.

As *possibilities scientists,* managers, teachers, and parents demand more of themselves. Much more! They require of themselves the conditions of their science:

- The relating, empowering, and freeing functions;

- The phenomenal modeling systems dedicated to discharging these functions;

- The interdependent processing systems that enable the phenomenal models to accomplish the freeing functions.

These are the conditions that make all phenomena live and grow fully. These are the conditions that empower the scientists to live and grow fully.

For the managers, organizations come alive! They are living, breathing, multidimensional, and asymmetrical—yes, asymmetrical—units, interdependently related to the organizations around them, and changing continuously, both internally and externally, in partnerships with others. They speak to managers from their own processing systems, *nested* in the marketplace positioning systems with which they are aligned, and housing human, information, and mechanical systems which they direct. They become for managers *"possibilities operations"* within which units and people and data and things are aligned with their essential missions: living, learning, working, growing, organizing, positioning.

For the teachers, content comes alive! Squanto really does relate to the Pilgrims in *their own "native language,"* and together, they collaborate to serve a survival function that we commemorate as Thanksgiving. Numbers really do relate to one another in discharging a social relating function as well as a mathematical

219

calculating function. Language really does serve a processing function as well as a communication function: it builds facts into concepts, and concepts into principles; it defines conceptual information in operational terms—functions, components, and processes; it transforms operations into multidimensional representations.

For the parents, children come alive! Yes, alive! Perhaps for the first time! Parents do not truly know their children until they relate to, empower, and release their children: this means relating to them internally, from the children's frames of reference; empowering them to realize their human potential; releasing them to find their own destinies. Healthy parents view their children in terms of the physical components that energize them, the emotional components that mobilize them, and the intellectual components that actualize them. Such parents enter the children's own images of living, learning, and working functions. They process interdependently with their children's own phenomenal experiences.

Possibilities scientists are biographers of phenomena. They live, learn, and work interdependently within, between, and among the phenomena they are studying. With this in mind, let us now summarize the definitions of (1) these phenomena, (2) the new possibilities science that addresses them, and (3) the possibilities scientists who represent this new science.

First of all are the phenomena as we have viewed them:

> **All phenomena are dedicated to socio-genetically determined, process-centric functions. All phenomena have multidimensionally formed, asymmetrically distributed, interdependently related, and continuously changing systems components dedicated to this process-centricity. All phenomena enable the changing systems components to accomplish the freeing social functions by interdependent phenomenal processing.**

In other words, the phenomena have a "mind" of their own: interdependent processing systems that enable them to mobilize their evolving components to achieve process-centricity. Thus the phenomena pilot *their own* destinies.

Humans can share in these phenomenal experiences through a science devised to surround their physical entities and penetrate their essence or souls. This is possibilities science, which has its own *phenomenal* definition:

> **The functions of possibilities science are to relate, empower, and free the phenomena. Its components are the phenomenal models dedicated to these freeing functions. Its processes are the interdependent human processes that enable us to generate these continuously changing phenomenal models.**

In other words, the science we employ to address phenomena is itself a phenomenon: it is defined by information capital development components; empowered by continuous and interdependent human processing; dedicated to discharging relating, empowering, and freeing functions. Possibilities science empowers us to share in the destinies of all phenomena.

Now we turn to the scientists who wish to share in this incredible journey in phenomenal experience. The possibilities scientists have their own *phenomenal* definition:

> **The functions of the possibilities scientists are to model phenomenal information. Their components are the interdependent processing systems that enable them to model information. Their processes are the operationalizing processes that empower them to model information.**

In other words, possibilities scientists are themselves phenomena: human processing components dedicated to information-modeling functions, empowered by mechanical processing operations. As such, possibilities scientists may share and even influence phenomenal destinies. In effect, the possibilities scientists are their interdependent processing systems: continuously engaged, modeling-focused, and operationally supported.

Concurrency ←→ The Matrix Solution

Remember concurrency? The possibilities processing paradigm! All of the projects illustrated thus far were products of this continuous interdependent processing: inductive, deductive, generative, interdependent, and possibilities processing.

Remember our original business mission? To generate new businesses from the concurrency model! At Carkhuff Thinking Systems, we implemented our design for concurrent processing. Our mission was to employ our breakthrough visions of science and our innovative technologies to generate new businesses (see Table 7-1). Again, we employed the GICCA paradigm to represent the market-cycle processing components of the market curve: generating, innovating, commercializing, commoditizing, and attenuating (or standardizing). We also employed the sources of new capital development (NCD) as the functions of new business: marketplace capital development (MCD), organizational capital development (OCD), human capital development (HCD), information capital development (ICD), and mechanical capital development (mCD). All of the operations now related interdependently and simultaneously in processing.

Again we employed the diagonal to represent *"the z dimension"* cells of our business positioning: new science, new technology, new business, new services, new products. All of the applications now related interdependently in processing. We labeled this approach *"The Matrix Solution."* We then summarized the *MCD* Matrix Solution operationally:

NCD functions are discharged by market-cycle components enabled by business-organization processes.

Table 7-1. Concurrency—The MCD Matrix Solution

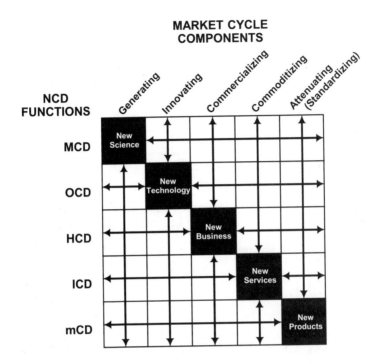

MARKET CYCLE COMPONENTS

We employed the Matrix Solution to process simultaneously and interdependently at every level of the organization. For example, we rotated our functions, components, and processes deductively, or counterclockwise, to develop an *OCD* Matrix Solution: new market-cycle functions achieved by new business-positioning components (see Table 7-2). Also, organizational processing systems enabling the new business components to accomplish the new market-cycle functions were introduced on *"the new z dimension"*: movement, partnership, people/teams, business models, and commercial initiatives. We then summarized the OCD Matrix Solution operationally:

Market-cycle functions are discharged by business-positioning components enabled by organizational processing systems.

Table 7-2. Concurrency—The OCD Matrix Solution

BUSINESS
POSITIONING COMPONENTS

MARKET CYCLE	New Science	New Technologies	New Business	New Services	New Products
Generating	**MOVEMENT** MOVEMENT - Vision - Mission - Scientific Srce. - Values - Brand				
Innovating		**PARTNERSHIP** PARTNERSHIP - Producer - Partners - Suppliers - Vendors - Customers			
Commercializing			**PEOPLE/ TEAMS** PEOPLE/TEAMS - Scientists - Board - Management - Development - Business		
Commoditizing				**BUSINESS MODEL** BUSINESS MODE - Preliminary - Private - "Angel" - Venture - Public	
Attenuating (Standardizing)					**COMMERCIAL INITIATIVES** COMMERCIAL INITIATIVES - Publications - Education - Products - Services - Solutions

Within these *"z dimension"* organizational processing systems, the following operations utilize the possibilities processing paradigm to process simultaneously and interdependently:

- *Movement*—Vision, Mission, Scientific Source, Values, Brand;

- *Partnership*—Producer, Partners, Suppliers, Vendors, Customers;

- *People/Teams*—Scientists, Board, Management, Development, Business;

- *Business Model*—Preliminary, Private, "Angel," Venture, Public Offerings;

- *Commercial Initiatives*—Publications, Education, Products, Services, Solutions.

These possibilities processing operations continue within all functions and components of the organization: operations are scaled, represented multidimensionally, and processed generatively.

Similarly, the internal organizational processes were rotated deductively to accomplish the following organizational processing objectives:

- *Business-movement functions are discharged by business-partnership components enabled by people/team processes.*

- *Business-partnership functions are discharged by people/team components enabled by business-modeling processes.*

- *People/team functions are discharged by business-modeling components enabled by commercial-initiative processes.*

- *Business-modeling functions are discharged by commercial-initiative components enabled by business-positioning processes.*

- *Commercial-initiative functions are discharged by business-movement components enabled by business-partnership processes.*

As such organizational processing continues interdependently, the Matrix Solution becomes *"the new SOP": "the new solutions operating procedures,"* or, better yet, *"the new science of possibilities."* Basically, a core concurrent-processing design yields the appropriate technologies addressing the appropriate objectives within, between, and among organizational processing systems: MCD, OCD, HCD, ICD, mCD.

For example, a full IT solution would address solutions at the following levels:

- Process,
- Application,
- Data,
- Network,
- Hardware.

In turn, a full Matrix Solution would directly impact the effectiveness and efficiency of the solutions at the following levels:

- CPU,
- Operating System,
- Hardware Connectivity,
- Network Routing,
- Data Modeling,
- Data Management,
- Programming Applications,
- Individuals and Teams,
- Enterprise and Inter-Enterprise,
- Object-Oriented Programming,
- Information and Process Objects,
- Graphics,
- Time Processes.

In other words, the processes of the Matrix Solution would extend to all human and organizational processing dimensions.

In short, one innovative design is available for simultaneous commercial applications. In the commercial world, the commercial objective is identified and the plan is developed. In the innovative process, the design is developed and the *interdependent process* emphasizes developing commercial opportunities.

In this context, the Matrix Solution optimizes the use of all of our technological tools:

- Mathematics,
- Geometry,
- Symbolics,
- Modeling,
- Language of processing.

To sum, the Matrix Solution optimizes performance within, between, and among all dimensions across all markets. The interdependent process-centricity of the Matrix Solution generates continuously elevating levels of effectiveness and efficiency for independent and planning-centric commercial enterprises.

By implication, concurrency yields *"a perpetual-motion business generator."* Its culminating outputs are found in continuously upgraded practices, products, services, solutions, and, yes, new businesses.

In fulfillment of our business mission, our process becomes our business: concurrency and the Matrix Solution become interdependent and synergistic processing partners. Concurrency generates continuously elevating environmental solutions. The Matrix Solution generates continuously elevating concurrent-processing environments. Our business becomes our process; our process becomes our business:

CONCURRENCY ⟵⟶ THE MATRIX SOLUTION

Paradigmetrics

"Paradigm" simply means model. *"Metrics"* simply means measurement. In this context, *"paradigmetrics"* means technologies that are dedicated to modeling and measurement: the measurement of modeling; the modeling of measurement.

In the new science of possibilities, all phenomena are modeled by their processing operations:

- Functions, or purposes;
- Components, or parts;
- Processes, or procedures.

These processing operations define the objectives of the phenomena:

> *Functions are discharged by components enabled by processes.*

Again, all operations are, themselves, processing systems.

In the new science of possibilities, measurements are also modeled by their operations: functions, components, processes. We label these measurement, *"standards."* When we are in a probabilities phase with parametric tools to produce products or deliver services, we dictate *standards of uniformity;* these reduce the variability or tolerance of production performance around some measure of central tendency. However, when we are in a possibilities phase with paradigmetric tools to generate breakthroughs and innovations and commercializations, we generate *standards of diversity;* these expand the variety and changeability of increasingly diverse performances.

We can view the relativity of modeling and measurement in all phenomena. For example, the standards may, themselves, become conditions that have their own functions, components, and processes. These newly generated conditions may, themselves, generate the functions of new phenomena, which, in turn, generate their own unique standards of possibilistic diversity and changeability and probabilistic uniformity. So standards may become conditions, and vice versa: all are phenomenal processing systems; all are relative or related; all are modeled; all are measured.

Indeed, the relationship between parametric and paradigmetric measurements is central to the differentiation of probabilities and possibilities sciences. Probabilities science utilizes parametric measurement of artificial, or man-made, phenomena: linear, independent, symmetrical, static, and planning-centric. Possibilities science uses paradigmetric measurement of all phenomena, virtual and naturalistic: multidimensional, interdependent, asymmetrical, changeable, and process-centric. In short, probabilities science forces the phenomena into a predictive *"box"* and controls their performance. Possibilities science empowers the phenomena to actualize their potential and frees their performance.

The profound implications of this simple definition of paradigmetrics cannot be avoided. Unlike parametric modeling, which moves by successive approximations toward static goals, paradigmetric modeling is continuously changing to reflect the accuracy of representations of phenomena. Correlationally, measurement is continuously changing to measure the accuracy of representations of phenomena.

To be sure, paradigmetric technologies are an entirely new scientific approach to both the measurement of modeling and the modeling of measurement. Like the new science of possibilities upon which they are based, these technologies make several powerful assumptions:

- That all dimensions of all phenomena are relative processing systems;

- That all phenomenal processing systems are modeled relatively;

- That all models of phenomenal processing systems are measured relatively.

The paradigmetric technologies are dedicated to these ends:

- All processing systems are *nested* in higher-order processing systems.

- All processing systems are rotated to become higher-order processing systems.

- All processing systems are impregnated with the socio-genetic code of higher-order processing systems.

In short, paradigmetric technologies are dedicated to multi-paradigmatic modeling and measurement.

Paradigmetric Modeling

Paradigmetric modeling, then, is dedicated to generating paradigms for both phenomena and their measurements. Indeed, the phenomena and their measurements are one and the same: the same scales that we generate for the phenomenal paradigms serve to measure the robustness of those paradigms. This book has provided several illustrations of interdependent common object modeling that serves a new organizing function—interdependent generative modeling—including electronic product definition (Chapter 4), hybrid modeling (Chapter 5), and paradigmatic modeling (Chapter 6). These generate *"possibilities objects"* for concurrent organizational modeling and process alignment.

One way to view paradigmetric modeling is as a horizontal assembly-line dedicated to developing products and services, solutions, or anything else. This view is represented in Figure 7-4. As may be noted, the process is initiated with the MCD Matrix:

- *MCD marketplace-positioning functions are discharged by components representing corporate capacities in MCD technologies.*

The process continues with the OCD Matrix:

- *OCD alignment functions are discharged by HCD human-processing components.*

Similarly, the process continues with the ICD Matrix:

- *HCD processing functions are discharged by ICD information-modeling components.*

Finally, the process culminates with the mCD Matrix:

- *ICD-modeling functions are discharged by mCD mechanical-tooling components.*

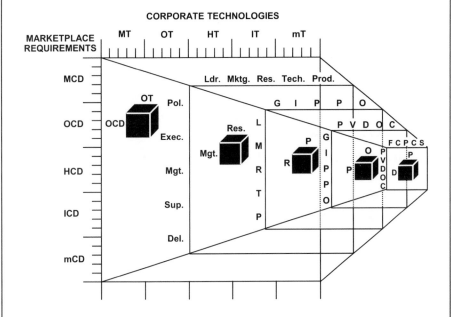

Figure 7-4. Paradigmetric Modeling

As shown, each successive NCD Matrix is *nested* within the previous matrix: MCD > OCD > HCD > ICD > mCD. This means that all of the processing occurs within the original MCD Matrix; therefore, all of the processing objects are generated within the MCD Matrix.

As also shown, all of the objects are sized according to the matrices that *nest* them; thus the objects diminish in size from MCD to mCD. The first, or largest, *nested* object occurs within the MCD Matrix:

- *OCD organization-alignment functions are discharged by OCD corporate-technological capacities.*

In turn, the next-largest object occurs within the OCD Matrix:

- *Management systems functions are discharged by resource-integration components.*

Similarly, the next *nested* object occurs within the HCD Matrix:

- *Resource-integration functions are discharged by processing components.*

Likewise, the next object occurs within the ICD Matrix:

- *Planning functions are discharged by operationalizing components.*

Finally, the last, or smallest, *nested* object occurs within the mCD Matrix:

- *Dimensionalizing functions are discharged by processing components.*

Again, all interdependent common object models are *nested* within the original MCD Matrix. We can employ paradigmetric modeling to generate any phenomena and their measurements.

Solid Process Modeling

A heuristic example for imaging paradigmetric technologies and their applications is *"solid modeling."* Just as solid product modeling dramatically increased productivity in designing and manufacturing components, *"solid process modeling"* will radically increase the productivity of the interrelated business processes of the extended enterprise.

The master model for solid process modeling is presented in the manufacturing illustration below (Figure 7-5). As we can see, the master model for manufacturing emphasizes market-player functions for involvement in constellation processing: producer, partners, suppliers, vendors, customers. In turn, the manufacturing processes are the components dedicated to discharging these market-player functions: feasibility, concept, definition, development, service. All other NCD systems are *nested* within this master manufacturing model: manufacturing processes discharging market-player processing functions.

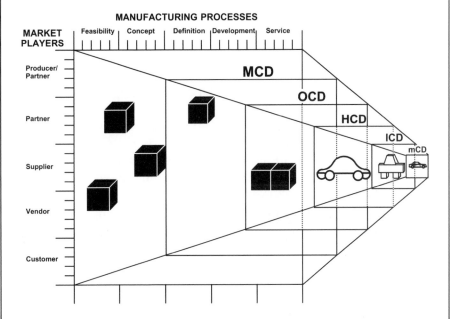

Figure 7-5. Paradigmetrics—Solid Process Modeling

In this context, all of the cells within the manufacturing model are interdependent common object models, with the solid process modeling system moving through common object modeling to the production of its product. Clearly, we can say that the contribution of paradigmetrics to solid processing modeling *is* common object modeling, or better yet, *"possibilities objects"* generated and implemented interdependently. They not only define the solid processing modeling system but also measure its potency in producing its products. And, of course, any interdependent common object modeling can be tailored to an industry's unique organizational requirements.

Again, probabilities science employs parametrics to plan its outcomes: its most potent practices yield minor variations of standardized products. Possibilities science employs paradigmetrics to process its virtual outcomes: its prepotent practices generate an infinite array of virtual, on-the-shelf products from which customers may freely choose the tailored options. With parametrics, we build products. With paradigmetrics, we build companies that build anything for anyone!

Furthermore, in using paradigmetrics, we are empowered to fulfill the destiny of information technology: information-modeling-driven manufacturing processes. Current organizational processing systems are out of sync, attempting to discharge MCD-positioning functions with mCD-tooling components. More powerful organizational processing systems will attempt to discharge MCD-positioning functions with ICD-modeling components. Future organizational processing systems will bring NCD operations into more appropriate alignment: first, to become driven by HCD processing; next, to become driven by OCD alignment; finally, to become driven by MCD positioning.

Of course, the rotations may be recursive: they may curve around to recycle iterations of rotations in a *"perpetual-motion business machine."* All that corporations require are the technological capacities of the paradigmetric tools of possibilities science so they can relate, empower, and free themselves to exponential and then potentially infinite corporate-productivity growth.

To sum, paradigmetric technologies draw upon the new science of possibilities to generate and measure paradigms for any phenomenon. Paradigmetrics are application tools of possibilities science: continuous modeling and measurement of continuously changing phenomena; continuous relating and empowerment to free phenomena to their potential so they can actualize their own changeable destinies. As a final note, these tools have been employed to create The Paradigmetrics Corporation, which was conceptualized, operationalized, dimensionalized, vectorialized, and phenomenally actualized with on-line concurrent processing.*

---

* The paradigmetrics business proposal was developed by Christopher Carkhuff, Alvin Cook, Bill O'Brien, Peter Rayson, and Darren Tisdale.

THE PARADIGMETRIC COMMUNITY

Paradigmetrics have much broader applications than solid process modeling. We may, for example, transfer our images to the evolving Internet community, as shown in Table 7-3. Note that NCD functions are discharged by community components:

- Business and industry, both on-line and off-line;
- Governance at all levels, on- and off-line;
- Science and higher education, on- and off-line;
- Schools and learning experiences, on- and off-line;
- Homes and neighborhood experiences, on- and off-line.

Table 7-3. The Paradigmetric Community Matrix

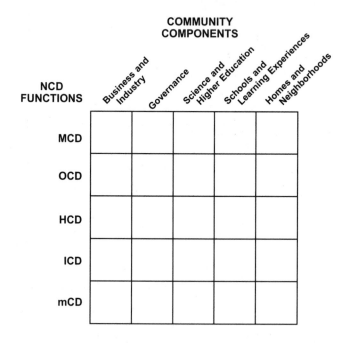

What the Paradigmetric Community Matrix conveys here is that interdependently related community components discharge NCD functions: *communities build new capital.* From another perspective, we may rotate these processing dimensions: *new capital development builds communities.*

Figure 7-6 illustrates how we may employ this matrix to expand business opportunities. As may be noted, we have already detailed solid process modeling as an alternative to mCD-driven manufacturing paradigms. We may now further ascend the NCD functions for other process-object-modeling business opportunities:

- *Information-process object modeling* for managing on-line businesses;

- *Human-process object modeling* for on-line business solutions;

- *Organization-process object modeling* for on-line business architecture;

- *Marketplace-process object modeling* for on-line creation of markets.

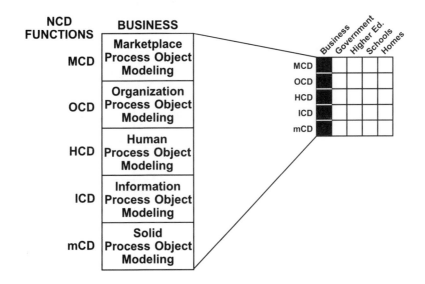

Figure 7-6. Business Opportunities in the Paradigmetric Community

What all of this means is that there are untold business opportunities in the paradigmetric community, both on- and off-line. Moreover, these businesses can contribute to community capital development, and vice versa: elevated community capital will contribute to economic capital development by the businesses.

In short, the *marketplace object modeling* that we are presenting sets into motion all other business modeling as well as community operations in a concurrent processing paradigm. We really can do good as we are doing well! Moreover, with the interdependent and collaborative requirements of the twenty-first-century global marketplace, we *cannot* do well if we *do not* do good!

We may also view the interdependent and synergistic modeling of the paradigmetric community in the latter's *source cells* (see Table 7-4). Here we find the following:

- *MCD-positioning functions are discharged by business and industry enabled by community-positioning processes.*

- *OCD alignment functions are discharged by governance enabled by community-alignment processes.*

- *HCD processing functions are discharged by science and higher education enabled by community processing.*

Table 7-4. Source Cells in the Paradigmetric Community

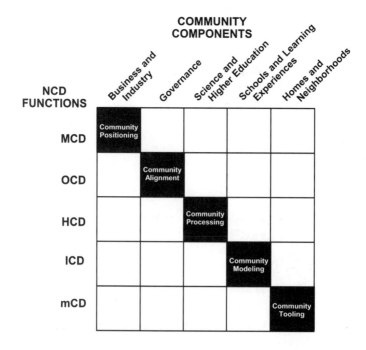

NCD FUNCTIONS	Business and Industry	Governance	Science and Higher Education	Schools and Learning Experiences	Homes and Neighborhoods
MCD	Community Positioning				
OCD		Community Alignment			
HCD			Community Processing		
ICD				Community Modeling	
mCD					Community Tooling

- *ICD-modeling functions are discharged by schooling and learning experiences enabled by community-modeling processes.*

- *MCD-tooling functions are discharged by homes and neighborhoods enabled by community tooling.*

Again, the enabling processes are illustrated on the diagonal representing *"the z dimension."*

If we position the community for community capital development (CCD), we may represent the entities as shown in Figure 7-7. Notice that the community components employ the following processes for discharging new capital development (NCD) functions:

- Community positioning for all entities;
- Community alignment of all entities;
- Community processing for all entities;
- Community modeling for all entities;
- Community tooling for all entities.

All components utilize community processes for discharging NCD functions.

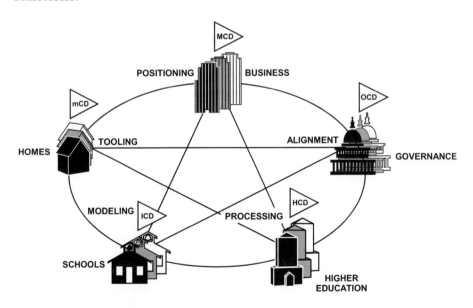

Figure 7-7. Community Capital Development

239

Cultural Capital Development: The Possibilities Community

Our efforts toward cultural capital development and the possibilities community began with a focus on education. Originally, we studied teachers who were exemplars and exemplary performers functioning at the highest levels within the probabilities symmetrical curve (see Figure 7-8). Our goal was to enable these exemplars to perform beyond the curve, at even more elevated levels of teaching. We pursued this goal by empowering the teachers to relate their students to their learning content. For example, we empowered exemplary performers in interpersonal relating skills, which enabled their students to perform at the top of the achievement curves: basically, the students would get nine or more months of achievement growth over the course of the school term, rather than an average of seven months or less.

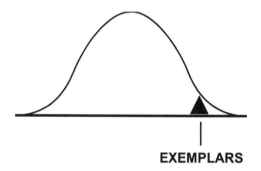

EXEMPLARS

Figure 7-8. Exemplary Performers in Probabilities Curve

At the same time, we studied the statistical exceptions: those who performed outside the probabilities curve (see Figure 7-9). Our goal was to discover and then explore the unique performance dimensions that enabled such individuals to perform beyond the actuarial curves. We analyzed, synthesized, and operationalized these dimensions in the course of our study. We found, for example, that a teacher might employ all kinds of tailored idiosyncratic teaching materials to personalize students' learning experiences; basically, the students would get two and a half years of achievement growth over the school term.

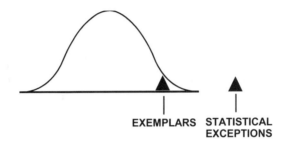

EXEMPLARS   STATISTICAL
EXCEPTIONS

**Figure 7-9. Statistical Exceptions Outside
the Probabilities Course**

One important result of our study of statistical exceptions was the evolution of the possibilities curve (see Figure 7-10). The presence of statistical exceptions initiated possibilities curves: they were asymmetrical, unlike the probabilities curve; they were also multidimensional, interdependent, process-centric, and changeable. In short, they defined new and continuously changing images of possibilities.

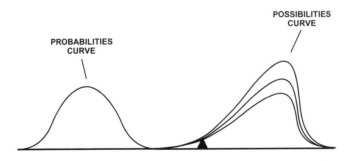

Figure 7-10. Evolution of Possibilities Curves

We may view the evolution of the possibilities community through the evolution from *"Probabilities Schools"* to *"Possibilities Schools"* (see Figure 7-11).* Initially, the schools were dedicated to S-R conditioned responding functions. In other words, the teachers wanted the students to make the correct responses to specific stimulus items. To this end, conceptual information components were dedicated. Through our work in interpersonal relating skills, we were able to empower the teachers to relate the learners to the conceptual information that enabled the correct responses to the stimuli. These *"Probabilities Schools"* dominated the Industrial Age through to the present time.

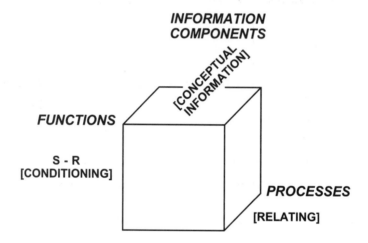

INFORMATION COMPONENTS

[CONCEPTUAL INFORMATION]

FUNCTIONS

S - R
[CONDITIONING]

PROCESSES

[RELATING]

Figure 7-11. The Probabilities Schools Operations

With the introduction of the Data Age, several different kinds of schools began to emerge, schools that were themselves statistical exceptions to the norm. We labeled such schools *"Possibilities Schools"* (see Figure 7-12). These schools understood the increasing need for discriminative learning they

---

* Schools project developed by Terry Bergeson, Andrew H. Griffin, and Shirley McCune.

were driven by S-O-R learning systems. In order to accomplish these functions, they dedicated operational information components. Finally, in order to enable the operational components to discharge S-O-R learning functions, they dedicated information representing processes.

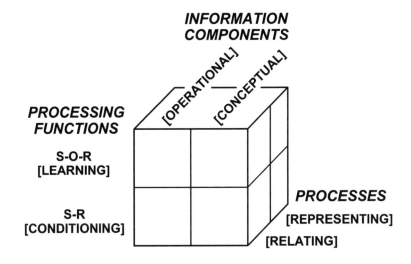

**Figure 7-12. The *Possibilities Schools*
Operations—Transitional Phase**

With the spiraling requirements of the evolving Information Age, the *Possibilities Schools* fulfilled their possibilities mission (see Figure 7-13): they were driven by S-P-R generative processing; they were powered by dimensional information components; they were enabled by reasoning processes. For educational purposes, we termed the learning processes *"The New 3Rs"*: relating, representing, reasoning. They were the skills required of our learner-products in the twenty-first-century global marketplace. They were the skills required to actualize the Information Age.

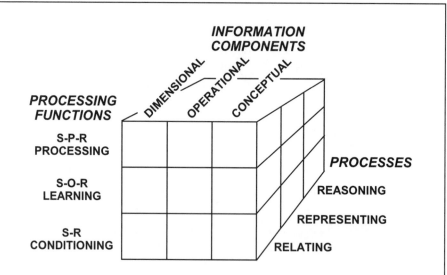

**Figure 7-13. The *Possibilities Schools*
Operations—Fulfillment Phase**

To put the *Possibilities Schools* systems in place, we required the *"Possibilities Education"* system (see Figure 7-14). This system is described as:

- Driven by S-OP-R organizational processing functions,

- Powered by vectorial information components,

- Enabled by interpersonal processing systems.

In addition, the *Possibilities Education* system needed to model the very functions that *Possibilities Schools* delivered:

- S-P-R generative processing,

- S-O-R discriminative learning, and

- S-R conditioned responding.

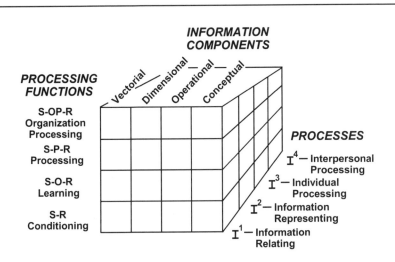

**Figure 7-14. The *Possibilities Education*
System Operations**

Finally, the *Possibilities Education* system culminated in the *"Possibilities Community"* system (see Figure 7-15). We described this system as:

- Driven by S-PP-R phenomenal processing systems of all kinds,

- Powered by phenomenal information components of all types,

- Enabled by interdependent phenomenal processing systems of all varieties.

As may be noted, we modeled the *Possibilities Community* system after the *"Possibilities Science"* system:

S-PP-R phenomenal processing functions discharged by phenomenal information components enabled by interdependent phenomenal processing systems.

All levels of all dimensions of possibilities science must be present in the *Possibilities Community*.

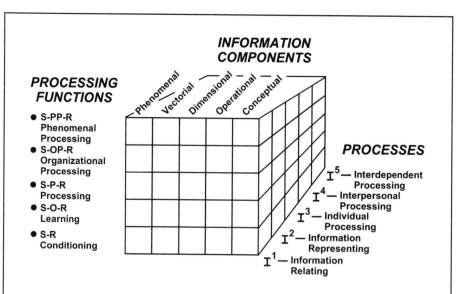

**Figure 7-15. The *Possibilities Community*
System Operations**

This, then, is the *Possibilities Community*[*] represented in its fullness to deliver possibilities processing functions: multidimensional, interdependent, asymmetrical, process-centric, changeable. As may be viewed in Figure 7-16, all entities carry primary responsibilities for processing objectives:

- ***The Home:***   S-R conditioned responding functions discharged by conceptual information components enabled by information relating processes.

- ***The Schools:***   S-O-R discriminative learning functions discharged by operational information components enabled by information representing processes.

---

[*] The *"Possibilities Community"* blueprint was developed with Andrew H. Griffin and Shirley McCune.

- *Higher Education:* S-P-R generative processing functions discharged by dimensional information components enabled by individual processing systems.

- *Governance:* S-OP-R organizational processing functions discharged by vectorial information components enabled by interpersonal processing systems.

- *Business:* S-PP-R phenomenal processing functions discharged by phenomenal information components enabled by interdependent processing systems.

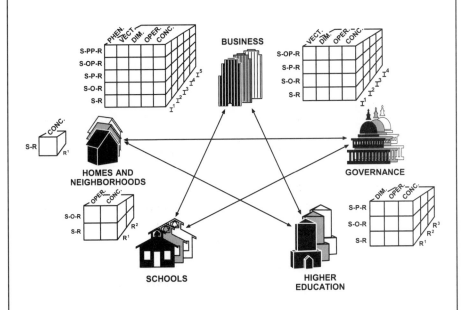

Figure 7-16. The *Possibilities Community* Operations

When the community is dedicated to the mission of cultural capital development, it relates to other communities and cultures (see Figure 7-17). At the highest levels, then, communities and cultures relate interdependently with other communities and cultures. This is mutual processing for mutual benefit, and it defines cultural capital development:

> *Interdependent relating discharged by free-enterprise economics enabled by free and direct democratic governance.*

In this context, cultural capital development is enabled by the paradigmetric community.

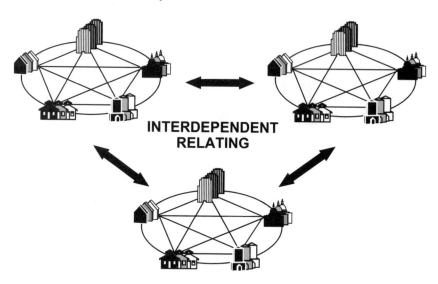

INTERDEPENDENT RELATING

Figure 7-17. Cultural Capital Development

In transition, the Internet community impels the world to embrace a global marketplace. This global community will be defined by elevated requirements for leadership *and* participation:

- Interdependent cultural relating,
- Free-enterprise economics,
- Free and direct democratic governance.

These dimensions define the healthy society of the twenty-first century and beyond:

Interdependent relating discharged by free enterprise enabled by free governance.

The paradigmetric community itself propels the Internet community toward cultural capital development empowered by new capital development systems:

- Community alignment to community positioning,
- Community processing for community alignment,
- Community modeling for community processing,
- Community tooling for community modeling.

All are accomplished by process object modeling on the Internet. We really can get there from here!

THE NEW SCIENCE OF PHENOMENAL POSSIBILITIES

This, then, is the new science of phenomenal possibilities. It does not reject the equations of probabilities science; rather, it views them as finite windows in space and time. Possibilities science seeks only to drive probabilities by possibilities, to give them perspective on what they do, to give them meaning where there is none.

Possibilities science has three fundamental process-centric principles:

- All phenomena are processing systems.

- All phenomenal processing systems are unequal in their power.

- All unequal phenomenal processing systems process interdependently.

We find the first principle in the functions of possibilities science: all phenomena are processing systems. All phenomenal processing systems transform resource inputs into results outputs and recycle feedback to improve and change performance.

249

We find the second principle in the components of possibilities science: all phenomena are unequal in their modeling power. All modeled phenomenal processing systems are related by their outputs and inputs: one system's outputs become another system's inputs.

We find the third principle in the processes of possibilities science: all phenomena engage in continuous interdependent processing. All phenomenal processing systems are enabled by their inequality to process interdependently: the inequality enables differentiated contributions.

We find these phenomenal principles replicated in both the conditions and standards of phenomena: all phenomena, including conditions and standards, are operationally defined by functions, components, and processes. Anything that can be operationally defined can be modeled, can process, and can change.

There are a number of enabling corollaries to these principles of possibilities science:

- All processing systems are *nested* in higher-order processing systems.

- All processing systems are *rotated* to generate new higher-order systems.

- All processing systems transmit the genetic coding of higher-order systems.

We find the first corollary in phenomenal systems: all phenomenal systems are *nested* in other phenomenal systems. All processing systems are *nested* in higher-order processing systems.

We find the second corollary in the rotations of phenomenal systems: all phenomena may be *rotated* to become the functions of a delivery system. All processing systems may be *rotated* to become higher-order processing systems.

We find the third corollary in the transmissions of the phenomenal systems: all phenomenal systems transmit the genetic coding with which they are impregnated. All higher-order processing systems transmit their genetic coding to lower-order systems.

Proceeding deductively, then, all phenomenal systems are *nested* in other phenomenal systems; all vectorial systems are *nested* in other vectorial systems derived from phenomenal systems; all dimensional systems are *nested* in other dimensional systems derived from vectorial systems.

We find the phenomenal systems in God's universes of phenomenal possibilities:

> *Phenomenal life-cycle functions are discharged by multidimensional components enabled by inter-dependent processing systems derived from asymmetrically curvilinear and continuously changing conditions and generating diverse and changeable standards.*

We find the vectorial systems in humankind's discoveries of nature's universes of vectorial possibilities:

> *Vectorial directing functions are discharged by socio-genetic coding components enabled by transmitting processes derived from changeable socio-genetic sourcing systems and generating changeable socio-genetic standards.*

We find the dimensional systems in humankind's universes of dimensional possibilities:

> *Operational processing functions are discharged by operational processing components enabled by operational processing systems derived from operational and changeable processing conditions and standards.*

Until now, humankind has proceeded inductively by conceptualizing and operationalizing phenomena far beyond the reach of its brainpower. That is precisely why it has evolved so slowly. Now, humankind can proceed deductively to dimensionalize phenomena by modeling them. If we can model phenomena, then they will live for us: we can understand the relationships within, between, and among phenomena.

With applications in processing, we will conquer the vectorial systems that source our dimensional models. If phenomena can live for us, then we can conquer the substance of nature's vectorial systems, for they are but windows in space and time on directional forces in God's curvilinear universes.

With dedication to transfers in processing, we can begin to understand the phenomenal conditions of our vectorial systems. If we can conquer the substance of the vectorial systems, then we can begin to discriminate the multidimensional, interdependent, asymmetrical, and changeable processing systems of God's phenomenal universes.

Again, we cannot begin to process generatively without dimensional modeling, which enables us to generate a myriad of interactions and alternatives in processing. We can, of course, process more powerfully with vectorial systems and phenomenal systems.

If we process deductively from phenomenal or vectorial models that we have built inductively, then we can process most powerfully and efficiently. In generic form, the phenomenal systems source the vectorial systems that source the dimensional systems. This deductive modeling is the most powerful source of expanding alternatives in processing. These interdependent and integrated systems empower us to expand the alternatives in processing exponentially, if not infinitely. This systematic expansion is the source of our systematic, generative processing.

In short, we cannot process what we have not modeled; and we cannot model that to which we have not related. Relating enables us to define phenomena informationally. Representing enables us to model phenomena operationally. Reasoning enables us to process phenomena generatively. These are *"The New 3Rs"* of possibilities: relating, representing, reasoning.

In this context, the new science of possibilities, especially in its deductive processing form, is an accelerator of evolution for any and all phenomena. We can generate totally new initiatives in an afternoon, initiatives that formerly may have required millennia to evolve.

The new science of possibilities culminates in the dedication of all of our processing systems to releasing or freeing phenomena to seek their own changeable destinies. Possibilities science is empowered by intervening to enhance phenomenal potential to express itself fully. It all begins with relating.

Above all else, possibilities give probabilities meaning. The explosion of possibilities within, between, and among phenomena provides a profound context for the meaning of probabilities beyond the miniscule images of performance, productivity, and profit.

For probabilities, the functions of description, prediction, and control define a very small *"probabilities moment"* in the infinite universes of phenomena.

For possibilities, the functions of relating, empowering, and freeing empower us to someday explicate the meaning of all phenomena in God's multiple and changing universes.

In summary, the real power of possibilities processing lies in the *nesting* and *rotating* of processing systems. All dimensions are processing systems. All processing systems are *nested* in higher-order processing systems and, in turn, *nest* lower-order processing systems. All processing systems are *rotated* according to the requirements of the higher-order processing systems in which they are *nested*. All processing systems transmit the socio-genetic coding of higher-order processing systems.

All processing systems, including especially possibilities science and its processing, are identical in characteristics with the phenomena they address: multidimensional, interdependent, asymmetrical, changeable, process-centric. Paradoxically, they find their commonality in diversity, their relatedness in *"inequality,"* their curvilinearity in modeling, and their predictability in change-ability. Process-centric at their core, they find their spiraling achievements in concurrency and their new currency in information capital development.

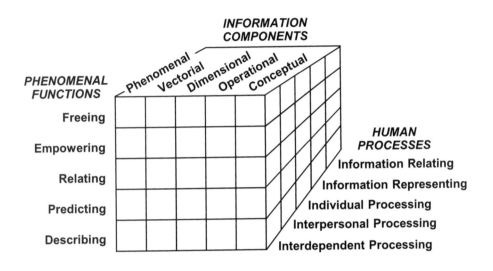

INFORMATION COMPONENTS

Phenomenal Vectorial Dimensional Operational Conceptual

PHENOMENAL FUNCTIONS

Freeing

Empowering

Relating

Predicting

Describing

HUMAN PROCESSES

Information Relating

Information Representing

Individual Processing

Interpersonal Processing

Interdependent Processing

The New Science of Possibilities

Selected Publications by Authors

The New Science of Possibilities. Volume I. Amherst, MA: HRD Press, 2000.

The Possibilities Leader. Amherst, MA: HRD Press, 2000.

The Possibilities Mind. Amherst, MA: HRD Press, 2000.

The Possibilities Organization. Amherst, MA: HRD Press, 2000.

Empowering. Amherst, MA: HRD Press, 1990.

The Age of the New Capitalism. Amherst, MA: HRD Press, 1988.

Human Processing and Human Productivity. Amherst, MA: HRD Press, 1986.

The Exemplar. Amherst, MA: HRD Press, 1984.

The Sources of Human Productivity. Amherst, MA: HRD Press, 1983.

Interpersonal Skills and Human Productivity. Amherst, MA: HRD Press, 1982.

Toward Actualizing Human Potential. Amherst, MA: HRD Press, 1981.

The Development of Human Resources. New York, NY: Holt, Rinehart & Winston, 1971.

Helping and Human Relations. Volumes I & II. New York, NY: Holt, Rinehart & Winston, 1969.

Acknowledgments

First, we would like to acknowledge the contributions of the core of research associates in Carkhuff Thinking Systems, Inc., who helped to develop some of the ideas presented in this work:

- Don Benoit, M.A., who contributed information representation operations,
- Chris Carkhuff, M.A. Cert., who innovated the organizational capital models,
- Alvin Cook, Ph.D., who built math models and coding systems,
- Barbara Emmert, Ph.D., who consulted on information systems development,
- Dave Meyers, M.A., who engineered organizational prototypes,
- Darren Tisdale, M.A., who innovated technological applications.

In addition, we owe a special debt to a number of people—themselves *"possibilities managers"* who made applications of our work at Human Technology, Inc.:

- John Cannon, Ph.D., Vice President, New Capital Development,
- Alex Douds, M.A., Director, Performance Systems Group,
- Sharon Fisher, M.A., Chief Operating Officer,
- Ted W. Friel, Ph.D., Information Technology Consultant,
- Richard Pierce, Ph.D., Director, Organizational Consulting Group.

We are particularly indebted to those scientists who contributed early on to our overall thinking:

- David N. Aspy, D.Ed., Carkhuff Institute,
- George Banks, D.Ed., Carkhuff Institute,
- David H. Berenson, Ph.D., Carkhuff Institute,
- Ralph Bierman, Ph.D., Carkhuff Institute,
- B. R. Bugelski, Ph.D., S.U.N.Y. at Buffalo,
- James Drasgow, Ph.D., S.U.N.Y. at Buffalo,
- Gerald Oliver, M.S., Carkhuff Institute,
- Flora N. Roebuck, D.Ed., Carkhuff Institute,
- Richard Sprinthall, Ph.D., American International College.

We also owe gratitude to pathfinders in business and industry who gave us opportunities to make applications:

- Rick Bellingham, Ph.D., Genzyme, Inc.,
- Russ Campanella, Genzyme, Inc.,
- Dave Champaign, Lotus Corp., IBM,
- Barry Cohen, Ph.D., Parametric Technology Corp.,
- John T. Kelly, M.A., IBM,
- Bill O'Brien, M.A., Parametric Technology Corp.,
- Russ Planitzer, Lazard, Inc.,
- Jack Riley, IBM,
- Peter Rayson, M.Sc., C. Eng., Parametric Technology Corp.,
- Carl Turner, General Electric,
- Norman Turner, General Electric.

We are also indebted to educational advisors with whom we processed interdependently to make extensive applications:

- Cheryl Aspy, D.Ed., University of Oklahoma,
- William Anthony, Ph.D., Boston University,
- Karen Banks, D.Ed., James Madison University,
- Sally Berenson, D.Ed., North Carolina State University,
- Terry Bergeson, Ph.D., Superintendent of Public Instruction, Washington,
- Mikal Cohen, Ph.D., Boston University,
- Andrew H. Griffin, D.Ed., Assistant Superintendent of Public Instruction, Washington,
- Shirley McCune, D.S.W., Director, Multi-State LINKS Project, Washington,
- Jeannette Tamagini, Ph.D., Rhode Island College.

Also, we express our gratitude to the trainers of Human Capital Development at the HRD Center, American International College, for piloting some of our work:

- Debbie Decker Anderson, D.Ed., Director,
- Cindy Littlefield, M.A., Associate Director,
- Susan Mackler, M.A., Holyoke Community College,
- Richard Muise, M.A., Assistant Director.

There are those who deserve our appreciation for their support in transforming these early manuscripts into readable books:

- Dave Burleigh, D B Associates for Marketing,
- Bob Carkhuff, HRD Press, for positioning,
- John Cannon, Ph.D., Human Technology, Inc., for his critical readings.
- Mary George, M.A., HRD Press, for editing.

Jean Miller deserves an exceptional note of recognition for implementing our *"rapid prototyping"* method of writing: about one dozen versions of each book were produced before final copy. Not only did she turn around high-quality typing, she also turned around high quality with timeliness. Not only did she generate creative graphics and layout, she also continuously retrieved lost files and, on at least two occasions, tracked down misdelivered manuscripts. These books are as much her books as ours!

Finally, we owe a debt of everlasting love and gratitude to those people who have been absolute in their commitment to enabling us to actualize our vision: our wives, Bernice and Gloria, who related to our experience, empowered our potential, and released us to the freedom of our scientific pursuits. For nearly 50 years, we have been saying, *"Give us another year and we'll get there."* Well, the *year* is up! And we got there!